Secrets of Korean
Alternative Medicine

Secrets of Korean Alternative Medicine

Amazing Stories of Healing Difficult Illnesses

Dr. Choong-Youl Oh

Secrets of Korean Alternative Medicine: Amazing Stories of Healing Difficult Illnesses

Copyright © 2011 Choong-Youl Oh. All rights reserved. No part of this book may be reproduced or retransmitted in any form or by any means without the written permission of the publisher.

Published by Wheatmark®
610 East Delano Street, Suite 104
Tucson, Arizona 85705 U.S.A.
www.wheatmark.com

Publisher's Cataloging-In-Publication Data
(Prepared by The Donohue Group, Inc.)

Oh, Choong-Youl.
  Secrets of Korean alternative medicine : amazing stories of healing difficult illnesses / Choong-Youl Oh.

  p. : ill. ; cm.

  Includes index.
  ISBN: 978-1-60494-530-0

  1. Hand--Acupuncture--Popular works. 2. Reflexology (Therapy)--Popular works. 3. Alternative medicine--Popular works. 4. Medicine, Korean--Popular works. I. Title.

RM184 .O43 2011
615.8/92   2011934406

Disclaimer

The entire contents of this book are based on research conducted by the author, unless otherwise noted. The publisher, the author, the distributors and bookstores present this information for educational purposes only.

This information is not intended to diagnose or prescribe for medical or psychological conditions nor to claim to prevent, treat, mitigate or cure such conditions. The author and the publisher are not making an attempt to recommend specific products as treatment of disease, and neither do they have any financial interest in the sale of the substances described in this book. In presenting this information, no attempt is being made to provide diagnosis, care, treatment or rehabilitation of individuals, or apply medical, mental health or human development principles, to provide diagnosing, treating, operating, or prescribing for any human disease, pain, injury, deformity or physical condition.

The information contained herein is not intended to replace a one-on-one relationship with a doctor or qualified health care professional. Therefore, the reader should be made aware that this information is not intended as medical advice, but rather a sharing of knowledge and information from the research and experience of the author.

The publisher and the author encourage you to make your own health care decisions based on your research and in partnership with a qualified health care professional. You and only you are responsible if you choose to do anything based on what you read.

Contents

Preface xi

1 Fifteen Years of Eczema Cured in One Treatment 1
2 Liver Cancer Completely Cured, Leaving No Trace 4
3 A Boy with Fourteen Holes in His Heart 9
4 A Man Who Called Me a Magician 11
5 Stomach Ulcer Cured 13
6 Curing Hepatitis and Heart Disease at the Same Time 18
7 A Woman Whose Heart Was Beating Too Fast 21
8 Instantly Stopping a Heart Attack 23
9 Myoma of the Uterus Cured Using Moxibustion 25
10 Healing Amenorrhea 29
11 Curing Acute Ovaritis 35
12 Curing Severe Rectal Bleeding 39
13 Treating My Former Student's Lower Backache 42
14 Instantly Curing a Lump 45
15 Curing Acute Lower Backache with a Single Treatment 47
16 Instantly Curing a Whiplash Injury 50
17 A Longtime Migraine Sufferer 53
18 Curing Grandma's Heavy Snoring 56
19 Curing My Brother, Who Became Mute after a Long Plane Ride .. 58

20	Story of a Grandma Who Sweats Severely While Eating.	62
21	Curing Severe Asthma	65
22	Instantly Stopping Severe Chest Pain after a Mastectomy	70
23	Curing Graves' Disease.	73
24	A Patient Who Had Heart Surgery	77
25	Bringing a Patient Out of a Coma	79
26	Curing a Patient with Parkinson's Disease.	83
27	A Patient with an Irregular Pulse	87
28	Relaxing a Cramped Hand.	89
29	Paraplegia Patient Walks after One Treatment	92
30	Instantly Curing a Person Disabled Due to Stiffened Fingers.	96
31	Story of a Patient with Chronic Esophageal Spasm	99
32	A Cramping Sole Instantly Relieved.	102
33	Curing an Injured Hip.	104
34	Treating a Woman Who Had Fallen from a Tree.	109
35	Curing Hemorrhoids.	113
36	Curing a Restless Child.	115
37	A Student Who Miraculously Escaped Renal Failure	119
38	Curing a Young Woman's Pimples	121
39	A Woman Suffering with Menstrual Pain	124
40	Curing a Patient Who Didn't Believe in Alternative Medicine	126
41	Discovering How to Stop a Nosebleed Instantly	129
42	Treating Eczema on the Scalp	132
43	Curing Sudden Ear Pain.	134
44	Curing Acute Tonsillitis	136
45	Realigning a Jaw That Was Causing Molar Pain.	138
46	Curing a Hockey Player's Dislocated Spine.	141
47	Curing Pancreatitis in a Simple Treatment	144
48	Curing a Lump on the Chest	147
49	A Seasick Young Man	150
50	Curing Severe Seasickness	154
51	Curing the Cold of a Visiting Overseas Speaker.	157
52	Curing a Chronic Headache.	160
53	Curing Chronic Esophagitis.	162
54	Curing a Minister with Chronic Sciatica	165
55	Curing a Patient Who Can't Stand Straight Because of Sciatica	170

56	Curing a Mounted Policewoman Who Had Often Fallen from Her Horse	174
57	A Patient Who Had Had Three Lower Back Surgeries	179
58	Curing Chronic Lumbago in a Truck Driver	183
59	Curing an Enlarged Prostate in One Treatment	187
60	A PSA Count that Dropped More than 50 Percent in One Treatment	191
61	A Miracle for a Liver Cancer Patient	194
62	A Stubborn Terminal Liver Cancer Patient	198
63	Instantly Raising a Friend Who Had Been Bedridden for Forty Days	203
64	Prostate Patient Returns for Second Treatment after Five Years	206
65	Reversing Hearing Loss	208
66	Korean Hand Therapy Explained	212
67	Korean Hand Therapy Applied	224

Appendix A: Life Extension Formula 229
Appendix B: Equipment and Materials 234
Appendix C: Frequently Used O-Chi Formulas 238
Appendix D: Meridian Charts 245
Index of Illnesses and Organs Affected 262

Preface

HUMAN LIFE MEANS LIVING WITH illness. Some illnesses are easy to cure, and some are not. Either way, we all suffer from such annoyances.

There are many different ways of curing illnesses. Take the problem of indigestion. There are many solutions, like taking digestives, undergoing massage, drinking soda, applying acupressure or acupuncture, using pellets, or simply chewing gum. They all work; the choice depends on the patient. One patient might have had a bad experience with side effects. For another patient, the problem might not have gone away quickly enough.

Modern medicine mostly prefers drugs and surgery. Some people are more comfortable with that route. At the same time, certain drugs are expensive, and some have unpleasant side effects. Some people fear surgery. For various reasons, some people might look for an alternative way to cure the problem.

Alternative medicine looks at illnesses from a different perspective than modern medicine does. Alternative medicine has its own method of finding the cause of and treating an illness. The most well-known alternative is Chinese medicine, which uses herbs, acupuncture, moxa, cupping, and acupressure. Many other folk remedies have developed through the centuries through trial and error.

Korean alternative medicine originates from well-known Korean-Chinese medicine. Dr. Tae-Woo Yoo developed Korean hand therapy in the early nineteen-seventies, and Dr. Chi-Kyung Kim developed Genesen Acutouch in the nineteen-nineties. These two alternative medicines invented by Koreans produce remarkably fast cures with very little pain.

As an undergraduate student, I studied pharmacy (BSc Pharmacy) and acquired a pharmacist's license at the same time that I acquired a teaching certificate for high school biology and chemistry. I taught chemistry at a Korean high school before being selected to train teachers at the university level in my audiovisual teaching methods. To pursue further studies, I came to Indiana University, where I earned a master's degree in education and a doctorate in media and technology. I taught at Indiana University and the University of Alberta (Canada). At retirement, I had taught for exactly forty years.

During my teaching career, I studied Chinese medicine, pulse reading, and various alternative medicines. By chance, I discovered and was captivated by Korean hand therapy, a near-perfect theory of systematically diagnosing and treating patients with effective results. I devoted myself to its study. In addition, I learned magnetic remedy, Genesen treatment, and electronic acupuncture treatment. I now have an established home clinic to treat patients.

My method is to find the cause of the illness and then apply the right treatment from my assorted skills. Using multiple remedies, I try to obtain the most effective result. (I never use acupuncture needles.)

Recently, I got a U.S. patent for a new method of curing whiplash injury and lumbago. Currently, I am applying for a patent for a new treatment for enlarged prostate.

This book illustrates my various treatment experiences—unique, amazing, and sometimes almost miraculous—written in short-story form. (To respect their privacy, all patients have been given pseudonyms.) Except for those treatment methods that have been patented, the treatments, including diagnosis, equipment, materials, and even formulas, I disclose in this book are described in enough detail that they should be relatively easy to follow. Consequently, readers can get similarly successful results.

I thank Margaret Sadler for assisting me in editing the English, as I translated from Korean.

For always encouraging me throughout my life and supporting my ideas, I thank my wife.

My wish is that all humans can live without unnecessary illness and achieve longevity.

Choong-Youl Oh, PhD
Edmonton, Canada
December 2010

Secrets of Korean Alternative Medicine

1

Fifteen Years of Eczema Cured in One Treatment

Every Wednesday for a year and a half, I visited a small aboriginal community about three hours northeast of Edmonton. Small houses are scattered around a big lake, and there is no doctor. There are fewer sanitary and other facilities than in the city. A well-equipped community center is used as a center for healthcare and cultural and sports activities. From time to time, the citizens invite various experts from outside to help people in the community, and so my wife and I were invited. A good-sized private room was designated as our clinic, with tables and chairs for simple treatments.

My office hours were 10:00 am to 4:00 pm, so we needed to pack up and leave the house by 6:00 am. Once we got on the highway, we had to drive straight east into the rising sun for two hours before turning north for another hour on country roads. At about the halfway point, we stopped to eat our breakfast in the car. Of course, we met the same problem driving home, this time with the setting sun shining straight into our eyes. After treating about ten patients, I would be tired and hungry, but there was no place to stop and eat until we were very close to Edmonton.

One day, Dorothy, a woman in her midsixties, came into the clinic. I said hello and extended my hand to shake hers. She responded, "No!" and hid her hands behind her. She said, "Don't touch me. I have eczema all over my body and it's terrible."

"How long have you been suffering with this?"

"About fifteen years or so."

"Why didn't you see a doctor?"

"I did! I saw many doctors, even a specialist at the university hospital. Every doctor prescribed a cream or ointment, but none of them worked. Some even made it worse, and it keeps spreading."

My general diagnostic test showed that energy from Dorothy's large intestine was excessive and her lungs were deactivated. I treated both hands with *lung-wet* formula. I told her not to touch the treatment for at least three days and ideally longer.

"Even if the eczema becomes very itchy, never scratch it. If it's unbearable, then wet a towel with ice-cold water to cool the area. Come back to see me next Wednesday."

The next Wednesday, Dorothy came in with both hands held out, saying, "Look what I have! Doctor, look here: it's all gone. All cured! This must be a miracle!"

"Congratulations," I answered, "I'm glad." This time, I shook her hand.

Then she handed me a full shopping bag.

"What's this?"

"This is all the medicine I received from the doctors. Do you want it?"

"No! You'll find a big garbage can on the way out."

Dorothy walked out happy.

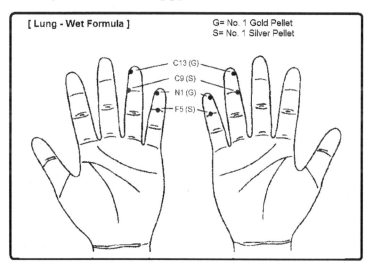

2

Liver Cancer Completely Cured, Leaving No Trace

One day I got a phone call from my younger sister in Los Angeles.

"Brother, I have a big problem."

"What's that?"

"Last week I went to see a doctor with a pain in my right chest. He diagnosed liver cancer."

"That is a problem. So what does your doctor suggest for treatment?"

"He wants to operate to take out a portion of my liver, and he wants to do it quickly. I'm to prepare for an operation as soon as possible. What should I do? Do you have any suggestions? I don't want to go through surgery."

"If you have that operation, you'll become very weak very quickly, and the recovery time will be long. Even though you have a portion of your liver removed, the statistics say that the possibility of survival is not good. Maybe you'll live another three or four years. I think you can try my alternative medicine. There is some expense to that, but at least you won't have to go through the operation."

She willingly agreed. "Tell me more."

"I'll get everything ready, and I'll be in LA in a week. In the meantime, don't eat or drink any strong spices or anything hot, salty, or fatty. Don't drink any alcohol or take any medicine."

"I'll be waiting for you next week."

I prepared the treatment plan, bought the necessary medicines and equipment, and headed to LA.

At the airport, my sister, her husband, and her daughter were waiting for me. We stopped at a restaurant to have lunch, and while we were waiting, I asked, "So how is your right chest?"

"It doesn't hurt too much; I just feel heavy and unpleasant."

I explained my treatment plan:

1. Do not eat hot, spicy, salty, or fatty foods; don't drink alcohol; don't take any other medicine.

2. Control your emotions; never get excited or angry.

3. Place pellets on both hands using the *liver-excessive* formula. Keep them there for three days, and on the fourth day, take everything off for a day. On

the fourth night at bedtime, put the pellets back on according to the *liver-excessive* formula, and leave them there for three more days.

4. Wear therapeutic rings on both thumbs.

5. Take the Chinese medicine *pien-tze-huang* (see page 7). Crush one tablet, and divide it into eight equal portions. Take one portion with warm water in the morning and another in the evening.

6. Prepare a special soup (see appendix A), and drink a cup each morning and evening.

After I explained this treatment, I gave her the following:

- two boxes of one hundred No. 1 silver pellets
- two boxes of one hundred No. 1 gold pellets
- two therapeutic rings
- twenty-four *pien-tze-huang* tablets (3 g each)
- recipe for special soup (see appendix A)

"You don't have to be scared of the cancer," I assured her. "You have to believe strongly that you can get rid of it. Stop everything else, and take care of yourself according to this treatment. Pray to God every day. Once every week or two, give me a call and report the results."

After I was home, she called me regularly and reported

that she was feeling better each week. About three months later, she said that she was feeling much better, and I felt that she sounded much more energetic. I asked her to visit her doctor and see how she was doing. "Yes, that's a good idea. I'll visit my doctor in a few days."

A week later, she called to say, "My Brother, I have good news. According to the X-ray, there's only something very small to be seen; it's smaller than a pea. The doctor said that if I continue this treatment, the cancer will disappear."

"That's very good news. Let's try another month. Do you still have *pien-tze-huang* left?"

"I finished it, but I'd like to try without it for awhile. If I need it, I'll give you a call."

After another month, her doctor ordered a CT scan, and no trace of cancer could be found on the scan. After that, the doctor told her, "Your liver cancer has disappeared without a trace."

I felt I did the right thing by advising against surgery.

Pien-tze-huang (traditional Chinese medicine) is a drug specifically for the treatment of hepatitis. It is made in Zhang Zhou, China. Each 3 gram tablet is packaged separately. To take it with the special soup for cancer treatment, crush one tablet into eight equal portions. Take each portion with warm water twice a day—morning and evening.

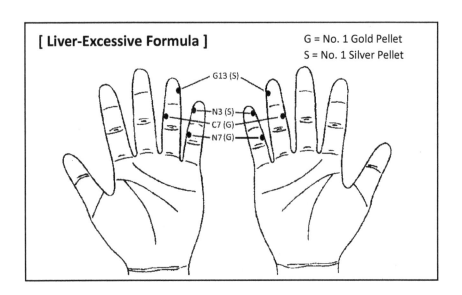

3

A Boy with Fourteen Holes in His Heart

A MOTHER BROUGHT HER ELEVEN-YEAR-OLD son, Kent, to my office. A cardiologist at the university hospital had diagnosed fourteen holes in Kent's heart. Ten of them were relatively small, but four were quite big. In his office, the doctor told the woman and her son that Kent likely would not live more than three weeks. Kent's mother was understandably surprised and upset because the doctor had said this directly to Kent.

His mother brought Kent to me for a second opinion. My diagnosis was that his heart was weak but overactive. My recommendation was to treat him by placing pellets on his fingers according to the *heart-excessive* formula and wearing therapeutic rings on each index finger.

I told Kent to keep the pellets in place for three days, and after three days, he could take them off. I also told him never to wear the rings to school in order to avoid teasing.

"Wash your hands when you get home from school, and put the rings on to wear throughout the night. The next morning, leave them at home again while you are at school."

About two months later, Kent's mother called me to report that Kent had regained strength very well. She wanted me to see him again. I checked him over and found that his heart was strong and almost normal. I told Kent's mother that she should go back to the cardiologist. "Don't say anything about my treatment, particularly about the rings."

About a week later, Kent's mother called. "The doctor says that all the holes but one have disappeared."

"That's great to hear, but until the heart is completely cured, continue to wear the rings."

About three years later, Kent came to visit me and asked me to treat his lower back. Kent had grown very tall and healthy. Now in ninth grade, he had sustained a lower back injury while playing hockey. Before I treated his back, I checked his pulse and found it perfect, which indicated a totally normal heart.

I'm always amazed at the power of the therapeutic rings.

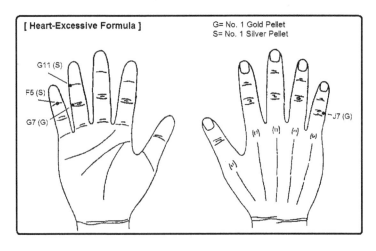

4

A Man Who Called Me a Magician

A TALL GENTLEMAN CAME INTO my office. "Hi, my name is Mike."

"How do you do? My name is Dr. Oh."

"Oh, you're the magician!"

"No, I'm not a magician."

"Well, everybody calls you that."

"I hadn't heard! But why are you here?"

"Please, lower my blood pressure."

"How old are you, and what kind of pills are you taking?"

"I'm seventy-seven, and I don't take any medicine."

I brought out my blood pressure monitor and took his blood pressure. The systolic was 182, and the diastolic 142. "Wow, that's quite high! Take off your shirt and lie down." A physical exam revealed that his heart was overactive and his small intestine was deprived. I used the *heart-excessive* formula to place the pellets on both of his hands, and I checked his blood pressure again. This time, the top number was 156, and the bottom was 110.

I put therapeutic rings on both index fingers and added one more No. 1 silver pellet at A16, located at the base of both middle fingers. I checked his blood pressure once again: 128 over 86. As soon as he saw those numbers, Mike yelled, "I'm right: you *are* a magician! No doubt about it!"

After he left, I was thinking, "Within twenty minutes, his blood pressure went from 182 to 128? That's remarkable. Maybe it's not completely wrong to be called a magician."

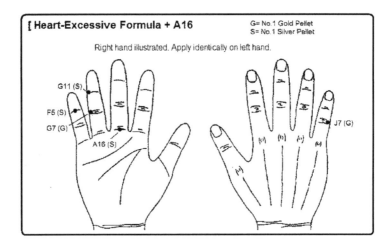

5

Stomach Ulcer Cured

Alice is a sixty-two-year-old living about three hours from Edmonton in a community of Ukrainian immigrants. She is a grandmother of a large family and is well respected in the community. Her daughter and grandchildren live in Edmonton.

Her granddaughter-in-law is the daughter of a Korean friend of mine. I was invited to their first son's first birthday party, where I was introduced to Alice. When I told her I was a practitioner of alternative medicine, she said, "I'm full of illness."

"Why do you say that?"

"I have about ten different problems. My whole body aches, I'm short of breath, and I can't walk properly."

"You must have one or two big problems besides the small ones. Which one is the worst?"

"First of all, I have a bad heart, a bad liver, and a stomach ulcer."

"Have you seen a doctor for these?"

"Yes, I went to the university hospital for a complete checkup. They found those three bad things. Other than that, I have a lower backache and aching knees."

"Wow, you have a lot of problems."

"Can you cure those things?"

"Yes, but I can't do everything at once. We'll have to do it one at a time."

"Doctor, when can we start?"

"I can start at the end of dinner. I'd like to fix your stomach ulcer first. It will take about an hour. Do you have time for that?"

"Oh, yes," she said.

I checked her pulse. I could tell her stomach was overactive, so I asked how long she had suffered from the stomach ulcer.

"More than five years."

"A stomach ulcer is not an easy illness, but there are many different ways of treating it. I have a special method that is quite effective, so I'll try that. The problem is that I must use a heat treatment called moxibustion. With this method, I have to light the moxa and place it on key spots, but this creates smoke and an odor, so we have to do it outside. The best place will be in the garage with the garage door open."

"Is it going to burn my skin?"

"As soon as you feel hot, I'll remove it from your skin so you don't get hurt."

In the garage, we opened the door, set up a table, and started treatment. I asked her to put both hands palm down on the table. I placed a small protective patch on the second joint of her middle finger on the left hand. I placed the moxa on top of that and lit it. The moxa started to burn down gradually and get close to the skin.

I asked, "Does it feel hot?"

"No."

"I need to continue until it feels hot."

After the first moxa burned down, I had to light a new one and repeat the procedure. Amazingly, she never felt hot. When the fifteenth moxa had burned down close to the skin, she said her hand was feeling a little hot.

When I positioned the sixteenth moxa, she said, "Oh, it's getting hot."

I removed it. On the right hand, when the eighteenth moxa was placed, she said it was getting warm, and a few seconds later, she said it was hot, so I removed it.

After that treatment, I used the *stomach-excessive* formula to place No. 1 silver pellets on both hands, plus the *basic* formula.

Two days later, Alice called me. "Doctor, my stomach feels much better. I always felt heavy there, but that has disappeared. I feel cured."

"Usually one treatment takes care of most patients, but your case is old, so be very careful. For the next week, eat soft, easily digested food. Avoid fatty and spicy foods."

"Doctor, I was doubtful of your treatment because you don't use any medicine. How can you treat stomach ulcers on your fingers? I'm very surprised by how effective your treatment was! Now that my stomach is cured, my heart is next. When can you treat that?"

I answered, "Don't rush. Your stomach has to be completely cured first, so we need to wait. Please call me when you've had ten days without your stomach giving you any trouble. Then, maybe we can say your stomach is completely cured."

She called ten days later to assure me that her stomach was completely cured.

It took a long time to develop this method, and I'm confident in treating ulcers this way.

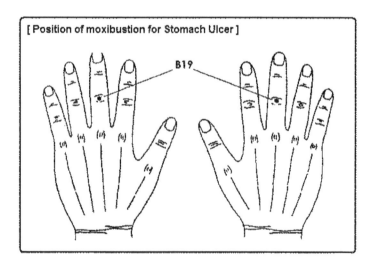

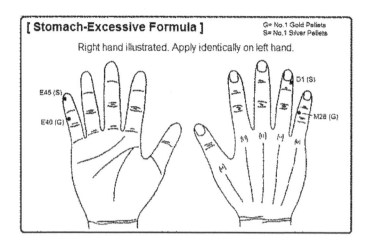

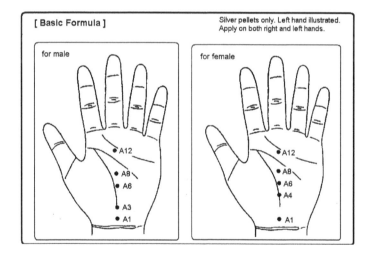

6

Curing Hepatitis and Heart Disease at the Same Time

Alice, whose stomach ulcer had previously been cured, asked me to cure her liver and heart conditions as I'd promised. On the examining table, I detected from her right pulse that her liver was overactive and her gallbladder was deactivated. On her left hand, I found that her heart was overactive and her small intestine was deactivated.

I started treatment on her right hand, placing pellets with the *liver-excessive* formula, and on her left hand, I placed pellets using the *heart-excessive* formula. I added extra No. 1 silver pellets at A1, A4, A6, A8, A12, and A16. I placed a therapeutic ring on her right thumb and another on her left index finger. I gave Alice the following instructions:

1. Keep the pellets in place for at least three days.

2. Both therapeutic rings should be worn twenty-four hours a day. If you have to remove them, put them back on as soon as possible.

3. Reduce fatty foods as much as possible.

4. Increase consumption of vegetables, fruits, and fish.

5. Walk at least thirty minutes a day at an ordinary, unhurried pace.

After three weeks, Alice's granddaughter-in-law called and said, "Dr. Oh, last weekend I visited Grandma, and she told me how she'd met an old friend at a shopping center a couple days earlier. Her friend was surprised to see her. 'I haven't seen you here for years. What brings you out now?' Grandma told her, 'I met a good doctor and my ulcer, hepatitis, and heart problem are all cured.' Her friend was really surprised."

"Do you think your grandma is getting better?" I asked.

"Yes, she's certainly different. In the past, whenever I visited her, she was always in her rocking chair. Now, she's working in the kitchen, and she goes out for walks."

"I'm happy to hear it."

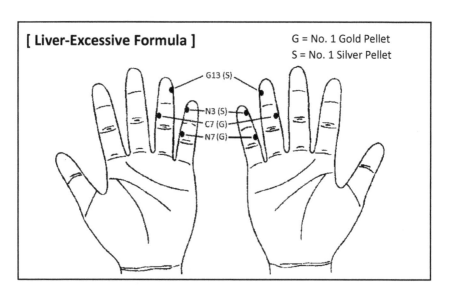

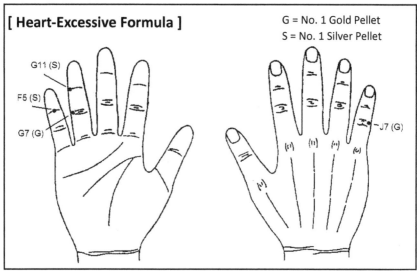

7

A Woman Whose Heart Was Beating Too Fast

One afternoon, I was resting after lunch when I got a phone call and heard a woman's trembling voice.

"Dr. Oh, it's Alice. Please come and help me. I'm in really big trouble. Please come quickly."

"Where are you?"

"I'm in Edmonton at my daughter's house."

"What's happened to you?"

"My heart rate is over 125 and I'm short of breath. My hands are trembling."

"Who measured your heart rate?"

"My daughter. She's a nurse at the university hospital."

"Yes, I see. I'll come right away."

I asked Alice's daughter for her address and left home with my treatment bag. When I arrived at the place, Alice was on the sofa, but she got up and came to me. I asked to see both of her hands and found that she had therapeutic rings on both thumbs and both index fingers.

I checked Alice's pulse, which was extremely fast. I got therapeutic rings from my bag and added them to both index fingers and asked her daughter to check her heart rate again. Her heart rate was 104. I put another ring on her left index finger, so she now had three therapeutic rings on her left index finger and two on the right index finger. I asked her daughter to take her pulse again, and this time, it was 90.

When Alice heard the pulse rate, both she and her daughter gave big sighs and looked very relieved. Her daughter exclaimed, "Wow! I've worked as a nurse for more than ten years, and I've never seen anyone's heart rate drop so fast."

I told Alice, "I don't know how you became so excited, but you need to learn to control your mind and try not to become so excited so easily."

The effect of the therapeutic rings always amazes me.

8

Instantly Stopping a Heart Attack

ONE EVENING, OUR CHURCH HELD a special session with a guest speaker. After the session, we got together in the fellowship room and enjoyed some refreshments. I stood visiting with the guest speaker.

All of a sudden, Mr. Han came from the kitchen toward me, both hands on his chest. His face was frozen in surprise, and he seemed to be enduring great pain, so I asked, "Are you in pain?"

He couldn't talk but nodded his head.

When I looked at his hands, I saw therapeutic rings on both index fingers. I approached him and removed the ring from his right index finger and put it on his left index finger, so he now had two rings on his left index finger.

Suddenly, Mr. Han exhaled a big breath and said, "Ah! Now I'm okay. Thank you."

He gave a little nod and walked back to the kitchen.

The speaker, who had been observing all this, asked, "What just happened?"

"He had a minor heart attack, but the danger is past. He's okay now."

The speaker was very surprised and gave me a weird look, staring at me for some time as if to say, *I've found a strange person here.*

9

Myoma of the Uterus Cured Using Moxibustion

A WOMAN IN HER THIRTIES came to my office from the Ukrainian community of St Paul.

"How are you? I'm Doris. I'm working in the government office in our city. About three months ago, I had a pain in my lower abdomen. I went to see a doctor, and he said I have a myoma. The doctor said he could cure me with a simple operation to remove my uterus."

I interrupted, "If you have that operation, there's no hope you'll have a child."

"Yes, and I'd like to have a baby, so I asked the doctor if that was my only choice. The doctor said yes, that was it, so I've come to get your opinion. Is that the only way to treat myoma?"

"There is another way to treat it, but it will take moxa and heat treatment."

"Moxa? What's that?"

I brought out some moxa, put a protective sheet on her finger, and placed the moxa on top of it. Without lighting the moxa, I showed her how if lit, it would burn down to the cardboard base. I explained that when needed, I would remove the moxa with tweezers, put on a new one, and keep repeating as necessary.

"You would repeat this every day at eight different spots on your hand. It requires a lot of effort and patience," I said.

"Doctor, if I do that, will the myoma be cured? How long will it take?"

"Certainly you'll be cured, but it might take a long time. Some people repeat the treatment immediately on the same spot, and that can save half the time. At the very least, it will take two to three months."

"Doctor, won't it burn my finger?"

"When the moxa is almost burned down to the skin, you remove it, and by doing so, you don't get burned; you just feel the heat."

Doris was thinking and looking at the moxa. "How much will it cost?"

"Moxa comes in boxes of two hundred, and you'll need at least five boxes." I told her the price of each box.

Doris thought for awhile and said, "Doctor, I'd like to think it over, and I'll need to talk with my husband. I'll give you a call in a few days. Thanks for the explanation."

After a week, Doris called and said, "Doctor, I decided to do the moxibustion, and I'd like to come and see you."

"I can see you tomorrow at eleven o'clock."

"Okay, I'll be there."

The next day, when Doris arrived, I put little marks on her hand showing where she needed to do the moxibustion: A3, A5, A8, F6, B1, B3, B5, and B7. I placed protective sheets on each spot and lit the moxa on each spot. I waited with tweezers in hand, and whenever she said it was hot, I removed the moxa and put it in the ashtray.

After all the moxa was burned, I asked her how she felt and if she had been burned.

"No, as soon as I felt hot, you removed the moxa, so I was never burned."

"The protective sheets are still on your hands. Shall I try once again?"

"Yes, please."

I put moxa on each of the protective sheets and repeated the treatment.

I gave her a drawing of a hand and marked the spots so she could remember them.

About three months later, I had a chance to pass through St Paul, so I stopped and called Doris. She came to meet me. As she approached, I thought she looked much better. On shaking her hand, I told her that she looked healthier.

"You must have gotten good results from the moxibustion. Are you still using it?"

"Yes, I feel healthier, but I used up the moxa about a week ago. Should I continue?"

"How is your lower abdominal pain? Has there been any change?"

"Some," she said. "I don't have pain anymore, but I'm still uncomfortable."

"I have some moxa with me. Why don't you continue?"

"Yes, I'd like to."

I gave her two more boxes of two hundred.

Two weeks later, Doris called. "Doctor, this morning I had a little bit of a lower backache, but shortly after, something like an egg-sized bloody lump came out."

"That's very good. Finally, you got the reward for all that time and effort."

"Thank you. I'm all cured, right?"

"Yes, I think you're completely cured now."

I was very happy to hear such results.

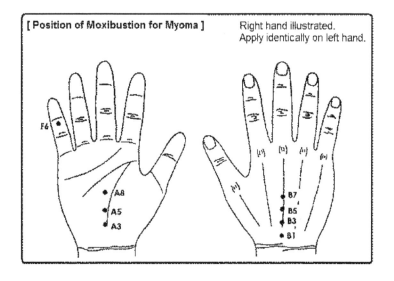

10

Healing Amenorrhea

Yusef is a Palestinian who immigrated to Canada with his family and has lived in Edmonton for more than ten years. In the past, he had come to me with heart disease, which I cured for him. One day, he called saying he wanted to book an appointment.

"What's the problem?"

He wouldn't say, saying only that he'd explain when he came.

Yusef came into the office with his daughter.

"Dr. Oh, this is my youngest daughter, Mariam. She's fifteen but she still hasn't started her menstruation although most of her classmates have."

She looked quite healthy, although a bit big for her age.

"Did you take her to a doctor?"

"Yes, we went to our family doctor and also a specialist.

They said there's no problem, she's very healthy, so there's no way to cure this."

"If so, I don't have any special way to cure her."

He looked a little worried.

"Doctor, if you can't fix her, then nobody can. You fixed my heart problem, and at that time, other doctors said they couldn't fix it. Please do something for my girl." Yusef was almost begging me to do something.

I told Mariam, "I'll give you a general exam today and check your state of health. Give me some time to study up and try to find some way to help you. Something must be wrong, and there must be some cause for it. Can you give me a week to figure it out? Is that okay with you?"

"Please do so." Before Mariam could finish her answer, Yusef was looking very happy.

I checked Mariam's pulse and found her to be in almost perfect health. She made an appointment to come back in a week.

Through various books, I tried to find possible causes and solutions to amenorrhea. Not many books even talked about it.

On the appointment day, Yusef and his wife came back with Mariam. I asked them if Mariam had ever been involved in a car accident or any other kind of accident. Yusef looked at his wife, and they both thought for a while, but finally said, "No, she's never been in an accident." I asked Mariam to lie facedown on the examining table. I checked her vertebrae from L1 to L4. From L2 to L4, she complained of a little pain when I pressed.

"Mariam, did you get your back hurt some time ago?"

"A long time ago, I slid down a slide at a park, and when I came to the end of the slide, I dropped hard onto the ground. That was when I was in first grade, so about nine years ago."

"Did it hurt very much?"

"Yes, I cried for some time."

"Okay, I think that accident is the cause of this whole problem."

When I said this, Yusef asked anxiously, "Doctor, can you fix it?"

"Now that I know the cause, there's a possibility of finding a cure."

I treated L2, L3, and L4 with Genesen for two minutes per vertebra, and afterward I placed No. 10 magnets on each vertebra. I placed No. 1 silver pellets on the four gynecology points—A4, A12, A16, and F6—and one therapeutic ring on each pinkie finger.

I told Mariam, "Keep the pellets on for three days, and wear the two rings twenty-four hours a day. You can remove the magnets after four days. Come back to see me after three weeks."

Six weeks later, Mariam returned to my office. She reported that there was no menstruation yet. I told her that she needed more frequent treatment.

"When does your winter vacation start?"

"It starts next week."

"That's good. Why don't you get treatment during winter vacation? Can you come twice a week?"

"Yes, I can."

"Okay, today is December 20. Can you come December 24, 27, 30, and January 3?"

"Yes, I can come then."

That day, I treated L2, L3, L4, and L5 with Genesen, two minutes each, and then placed No. 10 magnets on each point. On both hands, I placed pellets according to *kidney-deficient* formula and *stomach-deficient* formula. I asked her to continue to wear the rings on both pinkie fingers.

On the next appointment day (December 24), I treated L2, L3, L4, and L5 with Genesen for two minutes each and put No. 10 magnets on each point. On her right hand, I put No. 1 silver pellets using the *heart-deficient* formula, and on the left hand, I used the *spleen-excessive* formula.* Along the "I" meridian of both hands, I found the pain-sensitive corresponding points and treated them using twenty seconds of the electronic beam. I placed No. 1 silver pellets on each point on her hands.

On the next appointment day (December 27), I treated L2, L3, L4, and L5 with Genesen for one and a half minutes each and placed No. 10 magnets on each vertebra. Next to those points, one centimeter (about ½ inch) to either side of each vertebra, I treated with Genesen for forty-five seconds each. I placed No. 5 magnets on those points. Again, I found the pain-sensitive points on the "I" meridian on the back of her hands and treated them for twenty seconds each with the electronic beam. I placed No. 1 silver pellets on those points

---

* For *heart-deficient* formula and *spleen-excessive* formula, see appendix C.

on her hand as well as on A3, A4, A6, A8, A12, A16, and A18. That finished the treatment for the day.

The very next day (December 28), Yusef called. "Doctor, Mariam's menstruation started this morning. I thank you very much." I could hear the happiness in his voice.

"Yusef, thank you for the good news. Now at least one problem is over."

I was relieved because I had spent so much time and energy on this matter.

"Yusef, I'll cancel the next two appointments, but if you have any problems, don't hesitate to call. Please extend my congratulations to Mariam."

In the middle of March, Yusef called me again. "Doctor, Mariam's menstruation stopped after seven days in January. Now two months have passed and she has not had any further menstruation. I need to bring her to you again."

"Okay, tomorrow at four o'clock."

The next day, when Mariam came with her father, I thought she looked much taller. I felt that because her sexual organ function was normalized, she must have been getting the effects of female hormones.

"Mariam, are you well?"

"Yes, I am."

"So you are happy having menstruation, but wasn't it troublesome?"

"Yes, menstruation is cumbersome. I didn't know," she laughed.

"Then you have to experience such a bother for the next thirty years. Do you still want to take my treatment?"

"I have no choice. If I don't, there will be a lot of trouble at home." All three of us laughed.

That day's treatment was the same as on December 27.

Three months later, Mariam's sister came for treatment of tonsillitis, so I asked her about Mariam. She said that since the last treatment, Mariam had been menstruating regularly.

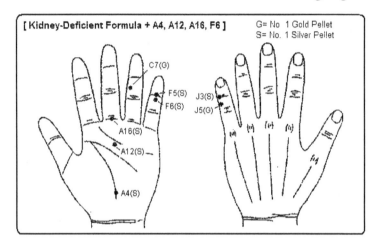

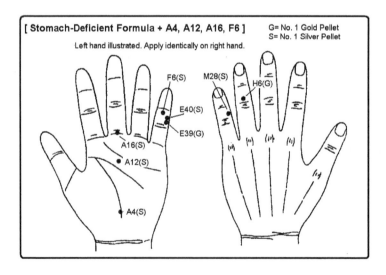

11

Curing Acute Ovaritis

I RECEIVED A PHONE CALL from my niece who lives in Tokyo.

"Uncle, I have a problem."

"What's that?"

"My lower left abdomen was painful so I saw a doctor, who diagnosed acute ovaritis. The doctor said I needed surgery as soon as possible. He said he would have to remove the ovary, but I'm not married and hope to have babies after I marry. If I have an ovary removed, won't there be a problem? Besides, I don't want to have a scar from an operation. Uncle, is there any other way to cure this problem?"

"Of course there's another way to treat this, but you have to come here if you want treatment from me. Try to do it very quickly. By the way, do you still have the rings I gave you when I was in Tokyo? Wear those on both pinkie fingers. Once your flight is set, let me know the details."

The next evening she called to say she'd be leaving in two days, and she gave me all the details.

When we met at the airport, I asked how her lower left abdomen was.

"Well, I just feel very heavy around that area, although not particularly sore."

"Do you have a fever? Do you feel hot?"

"Some, but not very much."

Her voice seemed very weak.

As soon as we arrived at home, before we even opened her luggage, I asked to see her at my clinic. I had her lie on the examining table. I touched the area, and I could feel a small lump under the lower left abdomen. I felt the warmth that indicates inflammation.

I treated her as follows:

1. At the point of inflammation, I set a blue Genesen perfectly perpendicular to her skin. I put the red Genesen about three centimeters (1 in.) toward the middle of her body and treated her for one minute.

2. I placed a No. 10 magnet against her skin at the point above the lump.

3. I asked her to lie on her stomach, and I treated L3 with Genesen for one minute.

4. I placed a No. 10 magnet on that point against the skin.

5. I told her to keep the therapeutic rings on her pinkie fingers continuously.

The next morning at the breakfast table, I asked how her lower abdomen was, and she said she had no pain anymore. She didn't know, though, if it was completely gone. She tried to touch the area while sitting there. I told her she wouldn't be able to find it through her clothes or while sitting. She stood and tried to find the lump but still couldn't. She looked puzzled.

"Good! You have nothing to worry about. Let's have breakfast, and then I'll examine you again."

She seemed comforted. After the meal, I asked her to lie on the examining table and touched the area, but through the magnet, I couldn't tell whether the lump was gone. I removed the magnet and tried to find the lump. I finally found it, but it was much smaller than the previous day, and there was little heat. I put the magnet back in place with new tape.

"Your lump is very small, so this means you're being healed. Just continue to keep the magnet there. It's not even been twenty-four hours, so forget about it because it's getting better."

"Uncle, thank you. I'm so happy that I don't have to go through surgery," she laughed happily.

The next morning I asked her again about her abdomen. She reported that she still felt no pain.

"This inflammation must have started with some bacteria. Do you know where that might have come from?"

"No, I don't know."

"Because you told me you have ovaritis, I have been thinking about why or where the bacteria entered. I've come to the conclusion that it may be from your bidet toilet [common in Japanese homes]. Most people who use the bidet feature use it after passing a stool. When you urinate, there's no problem, but if you use it after passing a stool, the shooting water might spread a tiny bit of stool to other areas. A small bit of stool might carry several thousand bacilli, which can easily be scattered. This is not as much of a problem for males because the distance between the anus and the urethra is greater, but in the case of a woman, the anus, vagina, and urethra are close together, so there's a greater possibility of infection. I suggest that after passing a stool that you always clean the anus with paper until the paper is clean, and then use the bidet feature. When you go home, tell the rest of your family, too."

Forty-eight hours after treating her, the lump and all of her symptoms had completely disappeared.

Because she had bought a ticket for a one-week stay, we traveled together to Banff National Park. When she left, she was so happy that all of her problems had been taken care of.

12

Curing Severe Rectal Bleeding

Alan, a handsome oil company employee in his thirties, visited my clinic. After we introduced ourselves, he said, "Doctor, I have a very strange problem."

"What is this strange problem?"

"Doctor, I'm not kidding, but I've talked to many doctors, and they all say this is very strange."

"Then they couldn't fix it?"

"Right, they don't know the problem, so apparently there's no way to correct it."

"Please explain your symptoms."

"About two weeks ago, after passing a stool, I noticed that the toilet was full of blood. Since then, every time I pass a stool, the toilet is the same. Losing so much blood, I lose my strength as well, but there's nowhere I feel pain. Someone told me, 'Go to Dr. Oh; he can find most problems!'"

First, I asked him to lie on the examining table. I tested his body and checked his pulse, but I found no abnormality. After asking him to lie facedown, I checked from L1 down to his tailbone, pressing each vertebra one at a time and watching for a reaction. At L5, his sacrum, and his coccyx, he reacted to pain. I checked carefully and discovered that his tailbone was broken and loose to the touch.

"Have you ever fallen on your tailbone while skiing or skating?"

"Yes, last winter while I was skating, I bumped into someone and sat down heavily on the ice. I remember the pain in my tailbone, but it disappeared after several days."

"At that time, you injured your L5, sacrum, and coccyx. This is the cause of the bleeding. Actually, the tailbone is pushed in and is unstable."

"Can you fix this?"

"Well, if you have a broken tailbone, we can't put a cast on it, so it's difficult to treat, but I'll try."

I treated him by using Genesen at L5, his sacrum, and his coccyx for one minute each and then placing No. 10 magnets on each bone.

"Alan, I've placed three magnets over that area. It might be bothersome, but I ask you to endure them for four days before you remove them. If you feel something unusual, please call."

"Yes, I'll do it. By the way, as people say, you are quite different!"

"Why is that?"

"Other people couldn't find the cause of the problem, but you did."

"You can say that after everything is cured."

"I'm pretty sure it's going to be completely healed," Alan said on his way out.

The next day, he called me. "Doctor, this morning when I went to the toilet, the amount of blood was really small compared to yesterday morning. It was only about a quarter of yesterday's. Am I getting better?"

"Yes, it is certain that you're getting a good effect. Would you please report to me again tomorrow?"

The next morning, he called again. "Doctor, this morning there was no blood. Am I alright now?"

"It seems so, but don't remove the magnets. As I said, keep them there for four days. After you remove them and there's no new bleeding, then you can be certain you're alright."

Three days later, he called to tell me there had been no bleeding since my treatment. He was very happy.

13

Treating My Former Student's Lower Backache

A FORMER GRADUATE STUDENT INVITED me to attend a small gathering over lunch held at a local Catholic retreat center. Most of the students had graduated five to ten years before, and most were well-established professionals. Some were in private business, and some were working in government and school systems. Because the graduation years were different, some were meeting for the first time, and because it had been such a long time for others, they decided to have everyone introduce themselves in turn.

I explained that since my retirement, I had learned alternative medicine—particularly Korean hand therapy—and was treating patients in my home office. Suddenly, everyone was interested in my new career and asked for more details. I explained the kinds of illnesses I treated, the treatments, and the results.

Richard asked, "Can you cure a lower backache?"

"Of course!"

"When can I visit your office?"

It was two o'clock on a beautiful May afternoon. Looking outside, I said, "I have my treatment equipment in the car. Why don't we meet at that bench under the tree after this meeting?"

"I would like that, but I have to undress, don't I?"

"This is hand therapy. I'll treat you using your hand. You won't have to undress!"

After the meeting, I met Richard on the bench and searched for the pain-corresponding points on the "I" meridians on both of his hands. I found reactions at I20, 21, and 22. I put two probes of the electronic beam on either side of his hand and treated each spot for twenty seconds. Then I placed No. 1 silver pellets on each point.

Richard said, "My back seems to be feeling better already. Thank you!"

Still, he looked doubtful because it had worked so fast. I suggested that he stand facing me about two meters (6 feet) away. I asked him to stand with his feet about thirty centimeters apart (1 ft.) and to raise both hands.

"Gradually bend from the waist until your hands are on your knees. Do you have any pain?"

"No."

"Now straighten up again and raise both hands to your

sides at shoulder height. Slowly turn toward the left about ninety degrees. How is your lower back? Any pain?"

"No."

"Twist to the right. How is your back?"

"No pain."

"Now come and sit on the bench with me."

Richard couldn't believe how instantly his lower back was relieved. He had injured his back five years before and had been suffering ever since.

"After such a long time with back pain, the pain will likely return. After three or four days, you'll need one or two more treatments from me." I gave him my business card.

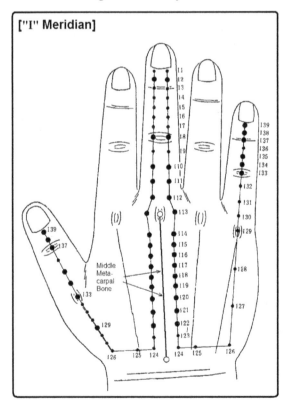

14

Instantly Curing a Lump

While I was treating Richard's lower back pain, a woman in her fifties came running out of the building to us.

"Doctor, would you please cure me? My name is Theresa, and I'm on staff here. I saw you curing Richard through the window."

"What kind of illness do you have?"

"This," she said, holding out her right hand. There was a 25 mm (1 in.) lump in the middle of the back of her hand, about 10 mm (1/2 in.) high. It looked hard.

"I'll look at you after Richard. Just wait a few more minutes."

When Theresa returned, I asked her to sit on the bench beside me and put her hand on her knee. I brought out the electronic beam unit and placed the black lead (–) at one side of the lump (lower end of the "B" meridian) and the other lead (+) on the other side of the lump (top end of the "B" meridian). I started measuring the time passed.

"Doctor, look! The lump is getting smaller!"

When she started yelling, I looked at the lump and saw that it was shrinking. In this case, I didn't need to watch the time, just the lump, until it had completely disappeared. It took about six minutes in total. It disappeared without any trace. Then, I placed a No. 6 silver pellet on the center of where the lump had been to finish the treatment.

As she left, she asked me to shake hands. As I shook her hand, I took the opportunity to feel where the lump had been. There was no trace of it, and the hand was beautifully healed. She said thank you repeatedly, and like a child, she skipped back into the building.

To me, it was almost miraculous. I sat there for awhile in thought.

15

Curing Acute Lower Backache with a Single Treatment

At about ten o'clock in the morning, a woman called. "Dr. Oh, would you please help my husband? He's in great trouble."

"What is the problem?"

"My husband got hurt at work yesterday, and he can't even move his foot. He has to go to work today but can't move. Can you please help him?"

"Can you come by at eleven o'clock?"

"We'll be there."

A middle-aged woman arrived with a man trailing behind her. Although he was shorter, he had very broad shoulders and a strong body, but he walked into my office like a powerless lamb. I invited him to sit, but he said he couldn't.

"Please explain how you got hurt."

"My name is Roy. I work for a construction company, and I specialize in carrying heavy loads of construction materials. Yesterday afternoon, I was carrying twelve pallets of bricks on a wooden board into a basement. As I walked down the stairs, I missed a step. I sat down heavily, and at that moment, I felt my lower back twist. I couldn't move. I couldn't lift the bricks, and I couldn't walk down the stairs. I called for help, and my coworkers moved the bricks one by one and removed the board. Still, I couldn't move. One of my colleagues came from behind me, and putting both hands under my arms, he lifted me gradually back up the stairs. I still couldn't move my feet. I leaned against the wall for awhile, and gradually, I could move. I came home in a friend's car."

I had to treat him as he stood. I used a probe to find the pain-corresponding points on the "I" meridian on both hands. I found two spots on each "I" meridian on either hand. Because the "I" meridian is U-shaped on both hands, I found a total of eight spots. I put a little mark with a pen on these points and then placed two leads of the electronic beam on the points across the middle metacarpal bone, treating him for thirty seconds. Because this allowed me to treat two points at a time, the treatment took two minutes in total. I placed No. 1 silver pellets on each point.

I asked him to stand up straight and spread his feet thirty centimeters (1 ft.) apart. I asked him to bring both hands up and then push back slightly.

"How is your back?"

"No pain."

"This time, bend forward until both hands touch your knees."

At that moment, the man shouted, "Ah, it's all gone! It's good!" For the first time, he smiled. "Thank you very much," and he extended his hand to shake mine. In his thick hand, I felt a very powerful man.

Roy and his wife were very happy as they left the office.

16

Instantly Curing a Whiplash Injury

Dr. Choi graduated from a British Columbia medical school and came to Edmonton for further study to become a specialist.

At about the same time, I had installed a Korean word processing program in my computer and was having problems using it. I needed help, so I asked for a Korean student who might be able to help me. Dr. Choi was the one who volunteered to help.

He came to my house one evening, and we began working together on the program. All of a sudden, he said, "Doctor, forgive me." He walked three or four meters (about 10 feet) away, raised both hands straight up above his head, and shook all over. He twisted his whole body and moved his head from side to side and up and down. I was surprised and didn't know what was going on. Two minutes later, he came back to my computer.

"What in the world were you doing?"

He told me his story. When he was a student in Vancouver, he was in a car accident and injured his neck quite badly. Since then, he had had to stand up and shake as he did every forty minutes.

"Is this a so-called whiplash injury?"

"Yes."

"Then how can you take classes? Do you do the same thing in the classroom?"

"No choice! At the beginning of the year, I explained it to the professor and my fellow students and asked them to understand if I stood up and shook in the middle of class. Now everyone expects me to do it."

"Dr. Choi, why don't you follow me?" I took him into my clinic. I asked him to sit on a chair and place both hands palm down on the table. I brought out the electronic beam and placed the black lead just below the middle fingernail at B27, and I placed the red lead on B19, the second joint. I treated him for forty-five seconds. I placed No. 1 silver pellets from B19 to B27 in a straight line about four millimeters (1/4 in.) apart. I treated both hands exactly the same way. He told me his neck seemed to be looser. He reached back to touch his neck.

The next day, Dr. Choi's wife called to thank me and said, "Doctor, you've fixed my husband! I thank you very much. He's no longer shaking himself."

Soon after, a woman called my office. "Dr. Oh, are you the one who cured my classmate Dr. Choi?"

"Yes, I am."

"This morning in class, he didn't shake himself, so all of the students were surprised. One of the girls asked why he wasn't shaking himself. He told us it was all cured. We wanted to know how. Then Dr. Choi described how you had fixed him."

She continued, "Doctor, can I see you myself?"

"Do you have a problem too?"

"Yes, I have been suffering from migraines for a long time. Can you please treat me?"

"Of course I can help you."

"When can I come?"

"How about tomorrow at three? Are you okay with that?"

"Yes, I'll be there at three o'clock."

So ended the conversation. Read on for a story of her problem with migraines.

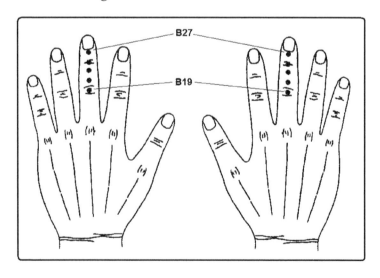

17

A Longtime Migraine Sufferer

The next day, as promised, Joanna, a thirty-five-year-old Latina, arrived at my office. After our greetings, I first asked about the course she and Dr. Choi were studying.

Joanna explained, "The course is in neurology, and the students all have their first medical degrees. Everyone wants to become a specialist. Dr. Choi wants to become a specialist in whiplash injury. I want to become a specialist in migraines because I have suffered from them for so long."

"So you're now studying an advanced degree after medical school. Do you know the cause of migraine headaches?"

"No, I don't. I've tried to find out. I've read many books, but I still can't find out what exactly causes migraines. I haven't found the right book yet. My main reason to take this neurology course is to get some help and hopefully find a solution."

"So, are you relying on painkillers? Is that your only way so far to endure migraines?"

"Yes. Do you know what causes migraine headaches?"

"Yes, I do. According to Korean hand therapy, which I practice, when you break the balance between the liver and the gallbladder, you get a migraine headache. You can also get a migraine if there's an imbalance between the stomach and the spleen. This theory is never understood by modern medicine. We look at human illness from a totally different perspective."

After I explained this much, I examined Joanna. Her present state showed that her liver was overactive and her gallbladder was deactivated.

"Your liver and gallbladder are out of balance. That's why you get migraines. The pain is probably either close to your eyes or at your temples."

"That's right! As a matter of fact, I always have pain at my right temple."

For treatment, I placed pellets according to the *liver-excessive* formula. I also put a therapeutic ring on both thumbs and told Joanna to keep the pellets in place for at least three days and to wear the rings twenty-four hours a day. Finally, I told her, "As long as you keep the rings on, you probably won't get a migraine."

"Thank you for a totally new experience, Doctor." Joanna left the office smiling.

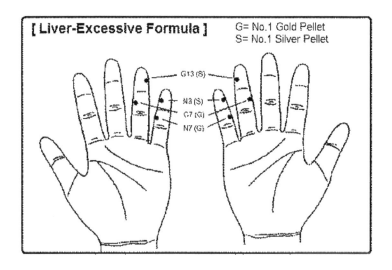

18

Curing Grandma's Heavy Snoring

GRANDMA CHA, EIGHTY-ONE YEARS OLD, immigrated to Canada about ten years ago. One day, I met her daughter at the shopping center, and she asked, rather embarrassed, "Doctor, can you, by any chance, fix snoring?"

"Yes, I have a way to do it. Is someone snoring heavily?"

"Yes, my mother snores so loudly at night that I can't get to sleep."

Grandma Cha's granddaughter interrupted, "She snores so loud that I feel like I'm going to fall over."

I was laughing, but the daughter was serious. "I'm not kidding. We have a terrible time coping with this. If you can solve it, please come to the house to help."

I went to the house the next day and met Grandma and her daughter.

I explained, "To control your snoring, you need to place

two No. 1 silver pellets on each hand at A8 and A28 each night before you go to bed. Place them on the same spot on both hands. If you forget before you go to bed and anyone hears you snoring, they can come to your bed and put the pellets in place on both hands without waking you up. Once the pellets are in place, within five minutes, the snoring will stop, and everyone can go back to sleep."

I asked her to extend her hands, and I put a little mark on each spot. "Those are the corresponding points for your nose. Place No. 1 silver pellets there before bedtime."

I put the four pellets in place and told her not to forget to put them there each night. I left her with a box of one hundred No. 1 silver pellets.

The next day, I got a phone call from her daughter. "Doctor, my mother didn't snore last night! Thanks to you, I could get some sleep. I didn't know you could fix snoring so easily. I want to say thank you for that."

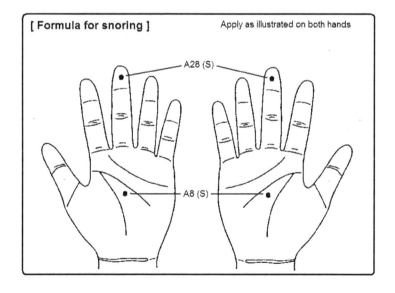

[Formula for snoring]   Apply as illustrated on both hands

A28 (S)

A8 (S)

19

―

Curing My Brother, Who Became Mute after a Long Plane Ride

WHEN THE PHONE RANG, I looked at the time: 6:40 AM. Who was calling at this time of day? I answered and heard, "Brother-in-law!" It was a familiar voice with a South Korean accent.

"Sister-in-law, where are you now?"

"I'm in Taegu."

"Good, you arrived safely."

"Yes, but we have a problem. Suddenly, your brother can't speak. We've admitted him to a Korean hospital."

"Oh, that is a problem. How did that happen?"

"We left Los Angeles last night and arrived in Seoul at noon today after the eleven-and-a-half-hour flight. We changed to a local plane and arrived shortly after in Taegu. At that time, your brother was complaining about a bit of a headache.

When we arrived home and met the family, he couldn't talk. We assumed he'd had a stroke, so we took him immediately to the emergency room at a Chinese medical facility. They treated him with acupuncture and then admitted him."

"So, are you calling from the hospital?"

"Yes, Brother-in-law. What should I do?"

"Has anyone drawn blood?"

"No, there's been no blood taken."

"What time is it there?"

"It's ten o'clock at night."

"I'll give you the phone number of a man who is practicing my kind of alternative medicine in Taegu, and you can ask him for help."

I gave her his name and phone number. About fifteen minutes later, she called back.

"Brother-in-law, I asked him to come and help, but he says he won't come out at night, so he says to wait until tomorrow and call him then."

"Sister-in-law, do you think you can find a drugstore open at this time of night? Go and buy a syringe needle and alcohol swabs. Use the needle to puncture both my brother's middle fingers at the tip.

"When you puncture, don't go too close to the nail because that would hurt. Puncture about two or three millimeters (1/16 in.) down from the nail into the soft skin. Draw about ten drops of blood from each middle finger. Once you do that, give me a call. Don't be scared. Don't worry; everything will be okay. Stay calm, and do this carefully."

After the phone call, I prayed to God for help and also

tried to calm myself. About forty-five minutes later, my sister-in-law called again, and I asked what had happened.

"I did just what you told me and drew out exactly ten drops of blood from each middle finger. Then he started talking. Here, you can talk to him yourself."

"Brother," he said, "I'm sorry I gave you so much trouble. I think I'm okay now." My brother's voice sounded the same as ever.

"I'm happy to hear that. What happened," I explained, "is that while you were on the plane, you sat in a narrow seat for a long time in the same posture; therefore, the blood started to coagulate. That's the cause. On the way home on the plane, get up and walk around as much as you can every half hour or so to prevent the blood from coagulating. In particular, the blood coagulates behind your knees, so you need to walk frequently to prevent more problems. Now, it's nighttime, so have a good night's sleep, and there'll be no more trouble. Don't change your travel plans. Be sure to visit our parents' graves, and return safely."

Several months later, we went to LA and had a family get-together. Sister-in-law recalled the events in Taegu. "I was so scared! All the patients who were sharing the hospital room watched me calling someone, then taking the blood out, and then they heard my husband start talking. Everyone was surprised. Their first question was, 'Who did you call?' All I could answer was, 'My brother-in-law.'"

Finally, she said, "Brother-in-law, you're a great man!"

Curing My Brother, Who Became Mute after a Long Plane Ride

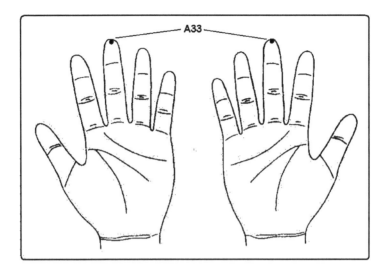

20

Story of a Grandma Who Sweats Severely While Eating

On a visit to Sao Paulo, Brazil, we stayed in a hotel. In the morning, we helped ourselves to our favorite foods at a breakfast buffet in the dining room. I went to our table and then went back to get coffee. As I tried to get coffee from the coffeemaker, someone spoke to me very kindly in Japanese, "Here, use it this way." I thanked her in Japanese. My table happened to be next to hers, so when I returned to my table, I noticed that she was eating alone.

"Why are you having breakfast alone? Are you staying here alone?" I asked quietly.

"No, my husband is here, but he can't come down for breakfast, so I'm eating alone."

She looked like she was in her mideighties. She was only eating bread and fruit, but her face was covered with sweat, and she was repeatedly wiping it with a paper napkin. I saw the sweat drop often onto her plate.

"Excuse me, but why are you sweating so much? Do you sweat like this every time you eat?"

"Yes," she said, raising her head to answer me.

"How long has this been going on?"

"I don't know, maybe four or five years."

"That's a lot of bother. I practice alternative medicine, and this is relatively easy to fix. Do you want me to cure it?"

"Really? Really, can you fix it? Many doctors I've asked can't fix it." She looked at me doubtfully.

"Yes, I can cure it relatively easily. I'm quite confident."

"Then why don't you try it?"

I gave her my business card and added my hotel room number. "Can you come around ten o'clock this morning?"

"Yes."

She arrived exactly at 10:00 AM. We sat at the small coffee table in the hotel room. I checked her pulse first. On the left side, I detected that her lungs were overactive, and on the right side, I could tell that her stomach was also overactive. On the left hand, I used the *lung-excessive* formula to place the pellets, and on the right hand, I used the *stomach-excessive* formula. I told her not to remove the pellets for forty-eight hours, and that finished the treatment.

She looked at me in disbelief because the treatment was so simple. It was only ten minutes altogether before she went back to her room.

The next morning when we returned to the dining room, the woman was already at breakfast. Watching her face,

I noticed no sweat, so I approached her and said, "Good morning. I don't see any sweat on your face!"

"It's really miraculous." She smiled broadly. "Thank you very much."

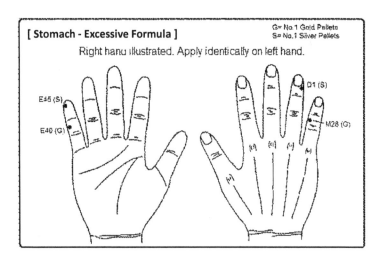

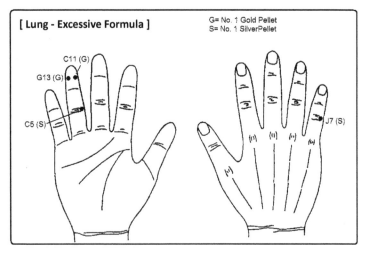

21

Curing Severe Asthma

ONE DAY, THE PHONE RANG. "Are you Dr. Oh? My name is Margaret. I'd like to make an appointment. Is that possible?" She asked for a particular date two weeks in the future.

"What kind of problem do you have?"

"I have severe asthma."

I told her that the date was possible and explained how to find my clinic.

On the appointment day, a woman in her thirties showed up. After our greetings, I asked her to write her name, address, phone number, and date of birth on my diagnostic sheet. I watched her fill out the form and noted that her address was in the Netherlands.

"Did you call me from the Netherlands?"

"No, I called you from Switzerland."

"What happened?"

She told me her story. She was a Dutch nurse at a big hospital in her hometown, but her asthma got so severe that it was impossible for her to work. On a medical leave, she went to a clinic in Switzerland for treatment. A few days before she had called me, she got a phone call from a relative in Canada. Her relative had strongly recommended that Margaret call Dr. Oh for an appointment, so she had called and booked a flight to Canada for a one-week stay.

"What kind of medicine are you taking now?"

She brought out a full page listing fourteen drugs.

"Are you taking all of these every day?"

"Yes, all except the last one."

The fourteenth was an emergency inhaler. I noticed that all of the drugs were very strong, and some were very dangerous medicines.

"While you're here under my care, can you stop using all of the drugs except the last one? If you have an asthma attack, you can use the emergency inhaler. Can you do it?"

"Yes, I'll try."

Through my examination, I detected that her right lung and her heart were overactive. On her right hand, I placed pellets according to the *lung-excessive* formula, and on her left hand, I placed pellets according to the *heart-excessive* formula.*
I added No. 1 silver pellets at A19, A20, A22, A24, B19, B22, and B24. I placed therapeutic rings on both index fingers and both ring fingers.

---

* For *lung-excessive* formula and *heart-excessive* formula, see appendix C.

I gave her instructions to leave all of the pellets and rings in place and to come back two days later. I asked her not to wear any other rings.

Two days later, she came for a second treatment. I asked how she had been doing during the past two days. "Did you cough?"

"No, I never coughed, and my lung capacity doubled." She appeared surprised.

"Are you carrying a lung capacity meter?"

"Yes, I have a portable meter. I couldn't believe you could get this much effect with one treatment without using any drugs."

"You need to change your idea of fixing disease with medicine only. There are so many other ways to cure illness. You don't always have to use medicine."

For the second treatment, I checked her pulse. There was not much change, although her pulse had slightly regained strength. I asked her to remove her top and lie facedown on the examining table. I treated C5, C6, and C7, plus T1, T2, T3, and T4 using thirty seconds of Genesen at each vertebra. Then, I placed No. 10 magnets on each of those points.

I replaced all of the pellets on her right hand according to the *lung-excessive* formula and did the same on her left hand according to the *heart-excessive* formula, but I omitted A19, A20, A22, A24, B19, B22, and B24. The therapeutic rings stayed in place.

Two days later, she came for the third treatment.

"Did you cough? Did you notice any change?"

"I never coughed. I feel a little more comfortable, and I feel some strength returning."

"That's good."

I checked her pulse. It was almost perfect, but I continued the treatment. I replaced all the pellets for the *heart-excessive* and *lung-excessive* formulas. I told her to keep the magnets on her back for another two days. On the hand meridian charts, I put a red mark at A19, A20, A22, A24, B19, B22, and B24 and gave her a box of one hundred No. 1 silver pellets.

"If your cough returns, place the pellets on those spots. Always keep the therapeutic rings in place."

So ended the treatments.

Margaret looked very happy. "Thank you very much. Goodbye, Doctor."

"Take my business card, Margaret, and if anything happens, give me a call."

About two months later, Margaret called me and said, "Doctor, please send me another therapeutic ring."

"What happened?"

"I haven't coughed since I left your office, but I started coughing last night, and then I noticed that my therapeutic ring from my left hand was missing."

I looked at her record: I had put a No. 7 ring on that ring finger.

"Yes, that's a No. 7 ring. I'll send it today; you can send me a check. By the way, how are you doing?"

"I'm back at work at the hospital, working happily every day. Thank you so much."

I feel very good about the result of this asthma treatment.

Curing Severe Asthma

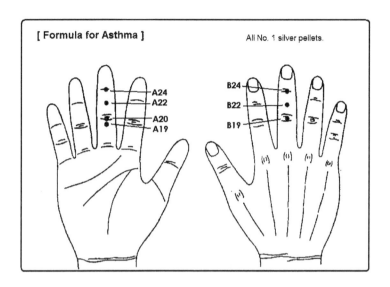

22

Instantly Stopping Severe Chest Pain after a Mastectomy

One night after supper, while I was finishing reading the newspaper, the phone rang. A male voice spoke.

"Hello, I'm Frank. My wife is suffering great pain. It's so bad that she's crying. I don't know what to do. I asked my Korean neighbor what to do, and he gave me your name and phone number. I know it's late, but could you come and help her, please?"

"Where does it hurt?"

"On her left chest just below the collarbone."

While we were talking, I could hear his wife crying in the background. I looked at the clock: 7:30 PM.

"I can be there by eight o'clock."

As a matter of fact, I didn't like to go out at night, but when I heard the woman crying, it was difficult to refuse.

When I arrived at the house, the husband and wife were sitting across the dining room table from each other. Her eyes were red and swollen.

"Where do you have pain?"

She didn't answer, just pointed about three centimeters (1 in.) below her collarbone.

"Is this the first time you've felt pain in this area? Do you know the cause?"

Her husband answered for her. "A year ago, she had a mastectomy after breast cancer. Since then, she's had several episodes of extreme pain in that area. We've gone to the emergency room before, and they've given her painkillers."

"How much does it hurt?"

"Very much. It's as if someone is stabbing my chest with a dagger. It feels like the end of the dagger is sticking out of my back."

"That's severe pain. Could you show me your left hand?"

She extended her hand. I asked her to put it on my left hand, and I used my right index finger to touch the C1 position at the base of her ring finger. I pressed there slightly. At that moment, she screamed, "Ah!"

The treatment was simple: I just placed a No. 1 silver pellet at that spot.

"How is your chest now?"

"Ah, I don't have pain anymore. It's all gone!"

The woman was surprised, but her husband was upset. "Were you faking?!"

His wife was flustered. "No! I was really in unbearable

pain. But how in the world could the pain go away so quickly and easily?"

Both she and her husband looked at the remaining pellets.

"What is this metal?"

"That's aluminum."

"Does this small piece of aluminum cure people?"

"Yes, if you place it in the right spot, you can see dramatic results."

"If we place one there if I have pain again, will it work?"

"Yes."

"Then I'd like to buy a box of that!"

She bought the whole box of No. 1 silver pellets.

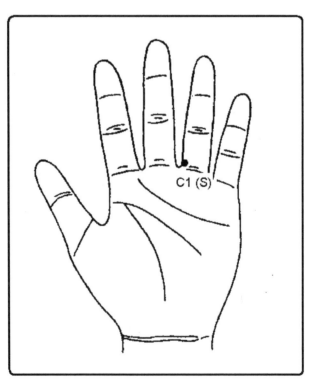

23

Curing Graves' Disease

In Canada, Thanksgiving Day is a national holiday on the second Monday of October. I had to visit the United States on business around this time. While I was getting ready for the trip, Frank (see chapter 22) came to see me.

"I understand you cure difficult illnesses. Today, I brought a very difficult disease."

"Are you carrying a disease with you? Let's see it!"

"No, no. It's not me. My sister has a very difficult illness and is suffering very much. I'm asking you to help her."

He explained that he was planning to visit his hometown, which was about fifteen hours from Edmonton, for Thanksgiving.

"My sister is so sick that she never comes out of her house. She has a swollen neck, and both eyes are sticking out. It's a horrible sight. Because her eyes stick out, everything is too bright for her. She can't watch television, and she can't leave the house in daylight. She's very temperamental and gets angry easily."

"Frank, that's called Graves' disease, and it's one of the most difficult diseases to cure."

"Doctor, can you fix it?"

"I don't know, but we have to try."

"But my sister can't come here."

"Then you have to do the work instead of me. The treatment your sister needs is different from the one I used to treat your wife. At that time, I used a No. 1 silver pellet. This treatment uses the same pellets, but because this is a more difficult disease, there are many more places to put the pellets. You have to be very precise and careful. I'll give you a chart, and you must follow my instructions exactly."

I gave him a chart with the formula marked on it.

"Wow, this is very complicated! You mean I have to place pellets in every place indicated?"

"Yes, and this is only one hand. You have to do both hands exactly the same."

"This is almost impossible! Besides, I can't cover all that with the pellets I have."

"You'll have to take a couple more boxes, but don't you think it will be worth it to cure your sister over the holiday? Once you put the pellets in place, don't remove them for three days. On the fourth day, remove everything, and give her a day's rest. The next day, repeat the placement, and leave them on for another three days."

"Wow, this is tricky, but I'll try!"

"I'll be visiting in the States for a week. Give me a call in a week and report the results."

I gave him two boxes of one hundred No. 1 silver pellets.

About ten days later, Frank called.

"It's Frank. When I went back to my hometown, I did what you explained. My sister started showing some effects from the second day on. On the fourth day, I removed all of the pellets, and on the fifth, I replaced everything. I had to come home on the sixth day. Last night, my sister called. Her eyes have gone back to their original position, but the swelling on her neck is still there. She can watch TV and knit again. She told me to thank you."

"That's good to hear. How do you feel about curing someone's illness?"

"Doctor, it's an unbelievably good feeling. I didn't realize the effect would come so quickly. I'd like to learn your Korean hand therapy."

"Frank, work hard, save money, and go to Korea to take the course."

"Doctor, is that possible?"

"Of course! There's a class taught in English for foreigners."

This was curing Graves' disease using someone else's hands. Getting such a good result at that distance surprised me.

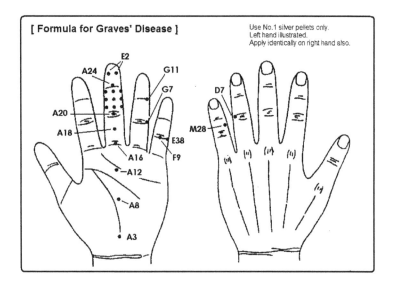

24

A Patient Who Had Heart Surgery

One day, an eighty-seven-year-old woman visited my clinic. Three years earlier, she had had a heart attack and bypass surgery. She thought that once she had surgery, her chest pain would disappear, but she still had a lot of pain. Also, she felt so weak that she couldn't do anything. She even had a difficult time climbing up onto the examining table.

In both hands, I detected that her heart was overactive and her small intestines were deactivated.

I treated her by placing pellets on both hands according to the *heart-excessive* formula. I put therapeutic rings on both of her index fingers. As soon as I put the rings on her, she said, "Doctor, my chest pain is disappearing!"

"Keep those rings on all the time, even while sleeping, but you can remove the pellets after three days."

About a month later, the woman called me and said, "Doctor, would you please send me another ring?"

"What happened?"

"I haven't had chest pain since I left your office, but last night, I felt chest pains again. Then, I discovered that the ring on my left hand was missing. I've searched all over the place, but I can't find it. Can you please send me another?"

I checked her record and saw that I had given her a No. 10 ring.

"I'll send you a No. 10 replacement ring today."

"Thank you. I'll send you a check right away."

I mailed the ring that day.

Some time after, her son came to visit.

"I just visited my mother. We held a big family celebration for her ninetieth birthday."

"How is her health?"

"She's much better. She walks about three kilometers (2 miles) every day, and she walks so fast that I can't keep up with her. Those rings are always on her fingers."

I was happy to hear this.

She lived to the age of 101.

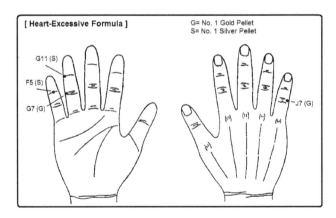

25

Bringing a Patient Out of a Coma

The Rev. Kwon is the minister of my church. One Sunday afternoon after the service, we were enjoying refreshments in the fellowship room. The Rev. Kwon approached me and asked to speak with me privately. We went to a quiet corner. "I have to go to Toronto on a flight tonight to see my family. I'm coming back by Saturday, so you don't have to worry about next Sunday's service. I just wanted you to know that I'll be absent." He looked unsettled.

I asked, "Is anything wrong?"

"Yes," he replied, and then he told his story. His ninety-year-old father had one leg amputated due to advanced diabetes. After the surgery, he fell into a coma and had been in a coma for more than two weeks. A few days before our talk, the doctor told Mr. Kwon's mother that he thought the final days were approaching quickly. He suggested that she bring all the family together for a final farewell.

The minister's elder sister came from Korea, his elder brother came from the States, and now he, the youngest, was traveling from Edmonton for the family gathering. As a minister, he was also mentally preparing to perform the funeral. He promised his mother that he would arrive by Monday morning by taking the Sunday-night flight.

Mr. Kwon told me, "I have a favor to ask."

"What's that?"

"By any chance—even for five minutes—can you bring back my father's consciousness so we can say goodbye and he can pass away in peace? Is that possible?"

"Mr. Kwon, I can't be certain, but you can try. Are you sure you want to try?"

"Yes, I will do anything for that opportunity."

I gave him a strip of ten No. 1 silver pellets and said, "Place one at H2 and one at I38." I put a small pen mark on his fingers to show the points.

"You must place them at exactly these spots. Put them in place as soon as you arrive at the hospital. Don't mention anything to anyone. Just put the pellets in place."

The next day, the Rev. Kwon called me.

"What happened?"

"My father woke up, but now he has loose bowels, so that's why I'm calling. Is there anything you can do to help with the diarrhea?"

"Any food you give him needs to be warm. Don't let him

have anything cold. Place something warm on his abdomen and keep it warm. Now, how many pellets do you have left?"

"I still have six."

"Okay, put one in the middle of his palms on both hands. That will work. By the way, congratulations on waking up your father. I'm very happy for you. I'm expecting you back on Saturday as planned."

After the Rev. Kwon returned, we had a chance to talk quietly. He told the story of his father's recovery. On that Monday morning after the plane had touched down, he went straight to the hospital from the airport.

"My father was in intensive care. I went close to him and placed the pellets on H2 and I38 on one hand and then in the same spots on the other hand, just as you said. A few moments later, my father opened his eyes. I approached and said, "Dad." He recognized me right away. Mother and the rest of the family in the room were so happy to be able to say, 'Go in peace.' Everyone was joyous.

"But my father asked, 'Go where? What do you mean by go?' We had to explain what had happened and that we had all gathered for his funeral. We all expected him to pass away and waited hour by hour throughout the morning and into the afternoon. He didn't die, and we soon realized that he would live. Each of his children left Toronto one by one."

As a matter of fact, this man lived for another six months. I thought the treatment would work, but at the same time, I was amazed that these two spots had had such a powerful effect.

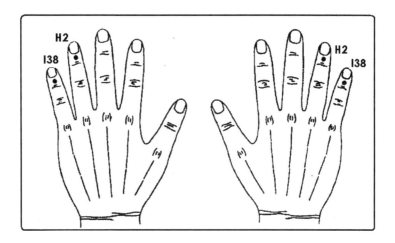

26

Curing a Patient with Parkinson's Disease

THIS HAPPENED WHILE WE WERE visiting the Canadian Native community northeast of Edmonton. One day, a sixty-four-year-old woman came to the clinic. Both of her hands were trembling. I asked, "Why are your hands trembling?"

"I have Parkinson's disease."

"Did you go to see a doctor?"

"Yes, I've been to many doctors and to a big hospital, but the diagnosis is always the same."

"What kind of treatment are you getting?"

"The doctors say there is no way to treat it, so they aren't doing anything. I hear you are different from other doctors, so I'm hoping you can help me. That's why I'm here today."

"How long have you been shaking like this?"

"Four years."

I wanted to know in what circumstances the shaking started. I understood that Parkinson's did not spread through a virus or bacteria. I asked, "Would you please tell me everything that happened in detail?"

"Yes, I'll tell you everything."

"Did you have any problem before this happened, or did it happen one day all of a sudden? Did you just wake up one morning to find your hands shaking?"

"Well, one day, some bad kids tried to beat up my son. I screamed at them and ran to grab my son away from them. At that time, my whole body started shaking, and that's when everything started."

I thought this was more likely coming from a psychological impact and therefore thought that her liver and heart might be sick. I examined her and checked her pulse. My examination revealed that her liver and heart were both overactive. I placed pellets on both of her hands according to both the *liver-excessive* formula and the *heart-excessive* formula. I gave her therapeutic rings for both of her thumbs and told her to wear them twenty-four hours a day.

I explained that her illness was not from bacteria or a virus, only from being overly frightened.

"You have to keep yourself calm and try to stay quiet every day. Living in a peaceful environment is essential to recovery. Don't forget to come back to see me next Wednesday."

The next Wednesday, I checked her again, but there was not much change in her condition. I gave her another set of therapeutic rings to wear on her index fingers. Now she had four rings—one on each thumb and one on each index finger.

I told her to eat less meat and more vegetables, fruit, and fish; not to drink any alcohol; and not to get angry or upset. I told her to return to the clinic in two weeks.

When she came back two weeks later, I asked if she had observed any change.

"The intensity seems to have lessened, but my hands are still shaking."

I could see that she shook somewhat less than before.

"Let's continue your lifestyle and see how it goes."

The last day of my visit to the community, the woman came to the clinic.

"Doctor, look!" She extended both of her hands, and neither hand was shaking.

"That is amazing!" I was so excited that I shook her hands.

"Doctor, really, I thank you very much. Have a good journey home."

I was so happy to see her joy-filled face.

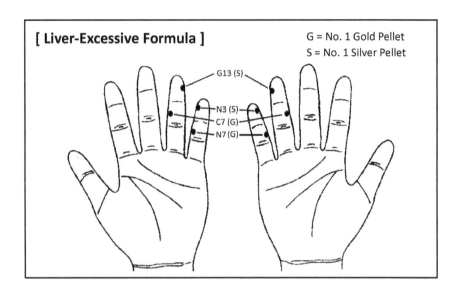

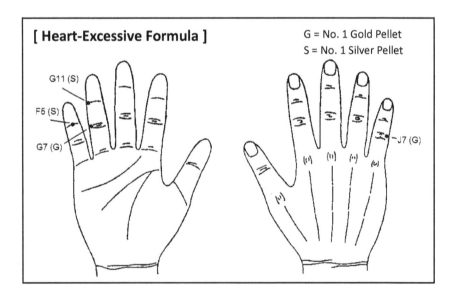

27

A Patient with an Irregular Pulse

Mr. Han is a Korean immigrant in Edmonton. We've been acquainted for quite awhile. Over the years, I've noticed that occasionally he coughs for no apparent reason. One day, I asked why he was coughing.

"Sometimes I feel that my heart has stopped beating. When I cough, it seems to start up again."

"Can I check you out?"

"Please."

I checked his pulse. While checking, I noticed that after four beats, it skipped a beat. Sometimes, I noticed two consecutive skips. When it skipped twice, that's when he felt he had to cough.

Through examining both of his hands, I diagnosed overactive heart symptoms. I gave him therapeutic rings for each index finger. As soon as I placed the rings on his fingers, he said he felt his chest area relax a bit.

Since then, he's always wearing the therapeutic rings, and his coughing is greatly reduced.

28

Relaxing a Cramped Hand

We were invited to attend my youngest sister's son's wedding in Philadelphia. At the rehearsal, I noticed the beautiful music of the pipe organ that the groom and bride would walk up the aisle to the next day. Sometimes, however, the music sounded a bit strange. It seemed that the organist was not too confident, but I didn't pay much attention. I was more focused on the greetings I was to give as the groom's uncle.

After the rehearsal, everyone gathered at a Chinese restaurant to share food and conversation. I was sitting across the table from the bride and groom. To my right was the bride's father. After the bride introduced her father and me, we talked to each other.

I commented that I had come a long way—from Canada. He had also come a long way; he presented his business card showing that he was from Indonesia. He explained that he ran a trading company there.

I introduced myself. "I practice alternative medicine."

The bride asked, "Uncle, what is alternative medicine?"

"There are many points on our hands that represent our whole body, and some of those key points heal problems in the body. The stimulation of those points makes people well and cures illness."

After I had explained that, the bride left the table and brought back a young lady from across the room.

"Uncle, would you please meet our organist? She's the wife of one of our ministers, and she's also the organist for our ceremony tomorrow. Recently, her left hand has been so cramped that she can't open it fully and therefore can't play easily. Can you fix that?"

"Of course I can. Don't worry about that. Let's have dinner first, and then I'll look at her."

After dinner, I brought the bride and the organist to a quiet corner and sat across the table from them. I asked the organist to show me both of her hands. She extended both hands, palms up. I asked her to close both of her hands into fists and then open them. Her left hand was unable to fully open; she could only open it halfway.

"What happens if you force it?"

"It hurts too much to do that."

I asked her to keep her left hand on the table, and using my thumb, I pushed down on K12. She pulled her hand away saying, "Ah, that hurts."

I brought out a No. 6 pellet and placed it on K12. I asked

her to open her left hand fully, and she was surprised at how easily she could do it. She quickly opened and closed her hand.

"Wow! It opens!"

"Does it still hurt?"

"No, it isn't hurting anymore! Thank you. Look, I can easily reach an octave."

The next day, after the wedding, the organist came to me smiling and said, "Thank you very much!"

> Note: This method works for any musicians who frequently use their fingers.

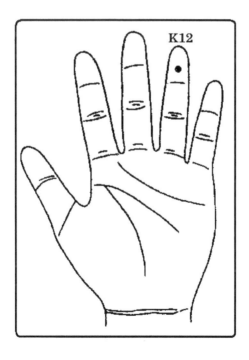

29

Paraplegia Patient Walks after One Treatment

The day after the wedding in Philadelphia, while we were having breakfast, my brother-in-law asked, very cautiously, "Brother, we haven't seen you for a long time, but I have a special request."

"What's that?"

"We have a friend who is suffering greatly with a problem. Her family belongs to our church. I hope you can cure her."

"What kind of illness?"

"We don't know the name. The problem is that she can't stand alone, she can't walk, and she can't stay sitting. They have gone to many doctors, but no one can come up with a name for the illness or a treatment."

"Was she originally a healthy person?"

"Mrs. Kim is in her late thirties and has two children in

elementary school. Mr. Kim has a small corner store and is trying to work hard to settle in as an immigrant. He volunteers in many different roles at the church."

"How long has she been suffering?"

"About three months or so."

After this conversation, I said that if they wanted to see me, I didn't mind.

My brother-in-law went to make a call and came back to report.

"They're going to New York to see a well-known specialist, so they're trying to get permission from the children's school principal to take the children out of school early."

"That's fine with me if they don't have time."

My brother-in-law was so concerned that he called again. After a long conversation, he finally came back.

"On the way to New York, they're going to stop and see you."

"What time are they coming?"

"About eleven o'clock. In the meantime, we have to go to work now. Would you please wait here and look after them when they come?"

I noticed that we still had an hour, so I readied my equipment. The time passed. Eleven o'clock came and went, and nobody showed up. I didn't like to wait.

Finally, the doorbell rang. I looked at the clock: 1:00 PM. I opened the front door, and a man was standing there with a woman on his back.

"Are you Mr. Kim? Come in, come in."

I asked them to come into the dining room. The woman

had both hands around her husband's neck, and he was pulling her across the room. Her feet left tracks in the carpet as they dragged across the room. I asked her to sit on one of the dining room chairs. Mr. Kim set her down and straightened her feet. He said she wouldn't be able to sit that way for long, no more than five minutes.

"That's okay; that's enough time."

I examined her and looked for the pain-sensitive points on the "I" meridian on the back of both of her hands.

Starting from the wrist and working toward the fingertips, I moved one millimeter (1/32 in.) at a time up her hand. I found three or four pain-sensitive corresponding points on each "I" meridian and marked them with a pen.

Placing the red lead of an electronic beam on the spot closest to the fingertip and the black lead on the closest spot to the wrist, I treated her for forty-five seconds. In the same manner, I did both branches of the "I" meridians on both hands.

"Now, can you stand up?"

When I asked, she looked at me strangely and didn't try to move.

I raised my voice a little. "Okay, I told you: stand up!"

She looked frightened and jumped up. At that moment she said, "What happened? I can stand!"

Then she looked at me. I cleared the way so she could move away from the table. I suggested that she walk. I held out my hand, and she took a couple of hesitant steps, but then she walked confidently all around the house—through the kitchen, the living room, and the dining room, laughing all the way. After she had taken three laps through these rooms, she came back to the dining room.

"Why don't you sit down so I can finish my treatment?"

She sat down again, and I put No. 1 silver pellets on each mark I had made on her "I" meridian. That finished the treatment.

Mr. Kim looked at this in astonishment. "Wow! Now my wife is recovered. I don't know what happened!" He rushed to the telephone and made four or five calls, repeating each time, "My wife is recovered!" He was so happy.

I said to myself, "Okay, I did another good thing today."

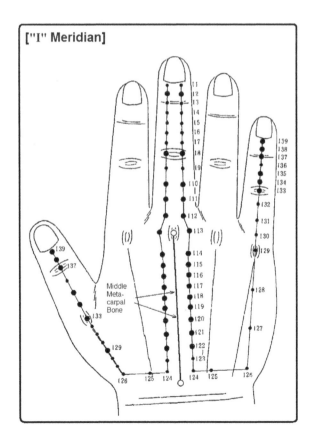

30

Instantly Curing a Person Disabled Due to Stiffened Fingers

Angela is a forty-three-year-old Caucasian office worker. I had cured her lower back pain about three years before. She called to make an appointment.

"I have stiff hands. Would you please see me?" It didn't sound very serious.

When she came to my office, her hands were all curled up, and she couldn't open or close them.

I asked her to open them wider, but she said it was too painful to do that.

"How long have you had this problem?"

"For about three months. At that time, they were hurting a bit, but now I can't move them at all."

"How about when you eat? Is it difficult to hold a fork and knife?"

"Yes. I can't do anything. I had to quit work, and then I had to sell my car because I couldn't make the monthly payments."

Her eyes filled with tears.

I asked her to put her right hand on the table. I touched the end of her ring finger. I pressed a little, and immediately she said, "It hurts!" I tried the same spot on her left hand, and she complained of pain there, too.

I brought out my Genesen and placed the blue Genesen on K13 and the red Genesen on K2, on the base of the ring finger. As soon as I touched the red Genesen to K2, she screamed, "Ah!" and stood up. I was surprised at her sudden reaction, so I removed the Genesen right away.

At that moment, Angela exclaimed, "I can move my hand freely!" She opened and closed her right hand fully.

She yelled again, "Look! I can do the same with my left hand."

She kept opening and closing both of her hands. She pulled her fingers back and shook both hands.

"Doctor, thank you! Now I can work again!" She started to laugh.

"Angela, this is a miracle. I treated only your right hand, but both hands are loose. You should thank God."

She repeated, "Thank you, thank you," and left happily.

Since then, quite a few patients have visited my office saying they had heard Angela's story.

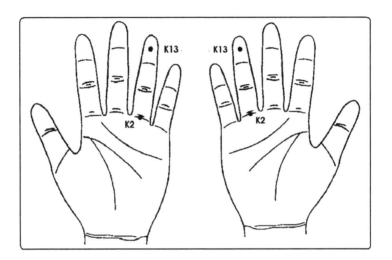

31

Story of a Patient with Chronic Esophageal Spasm

A YOUNG BELGIAN MAN IN his thirties came to see me. He gave a good physical impression, but his face looked dark.

"Doctor, my name is David, and I have brought you a difficult problem."

"What difficult problem is that?"

"I cannot eat."

"Is that why you look pale and thin?"

"Yes. Whatever I eat doesn't go down smoothly. I feel like it's stuck in the middle of my chest. Then, I feel pain in my chest and have to stop eating. I've consulted many doctors, and they call it esophageal spasm, but no matter how much I take the prescribed medicine, it doesn't get better."

"How long have you been like this?"

"Over eight years. I've taken many medicines, but they don't work. I called my mother in Belgium, and she suggested that I come home and see the family doctor who looked after me since I was a baby. I went back to Belgium last summer, but that doctor said the same thing as the other doctors, and he didn't know how to treat it."

"Then how do you maintain your life if you cannot eat much?"

"I can manage lukewarm soft foods or juice if I take it very slowly."

I asked him to take off his shirt, and I checked his body and his pulse. Overall, he was quite weak, and his pulse was weak, too. Otherwise, no big problem was evident. I asked him to lie facedown on the examining table, and I started to examine his back from C1 to T6. I pushed every vertebra to see if I got a painful response. The vertebrae C3, C4, C5, and C6 caused some slight reaction, and then I noticed that those four vertebrae were tilted to the right.

"David, have you ever been in a car accident?"

"No, why do you ask?"

"Four of your neck bones—the lower cervical—are tilted. You must have had some kind of impact to cause that. You don't remember a car accident?"

"No."

"Do you play soccer?"

"Yes, I've played soccer since I was at university. I still play with my company."

"Do you often perform headers?"

"Yes, one time I did a header while playing and had a lot

of pain. I stayed home for several days. I remember being very sore at that time."

"Okay, maybe that's the problem. You're ill because your cervical vertebrae have been damaged. If your C3, C4, C5, and C6 are not healthy, you can get stomach, esophageal, and sometimes diaphragm spasms, as well as heart and lung illnesses. I have to correct the tilted vertebrae. If I correct just that much, your esophagus problem will disappear automatically."

I treated C3, C4, C5, and C6 with Genesen for one minute each. Then, I placed No. 10 magnets on those vertebrae. I had him stand two meters (6 ft.) away facing me, and I asked him to turn his head slowly to the right and to the left.

"How does that feel?"

"Much easier than before."

I asked him to slowly look up and down, and I repeated, "How does that feel?"

"There's no pain. I can move my neck very easily."

While he was putting his shirt back on, I reminded him to keep the magnets on the vertebrae for at least four days and asked him to report back in about a week to tell me how he was feeling.

"Doctor, thank you. I'll give you a call."

A week later, David called.

"Doctor, I'm okay now. I can eat as I used to with no problem. Thank you very much."

I heard the relief and happiness in David's voice.

32

A Cramping Sole Instantly Relieved

One day, at a large shopping mall, I was walking rapidly toward my destination when suddenly, my right sole cramped up. I was in great pain. I couldn't put my right foot down, so I limped awkwardly on my heel, finally leaning against a wall for relief.

I took a No. 1 silver pellet from my pocket and placed it on the end of my right pinkie finger at F3. About five seconds later, my right sole relaxed completely, and I could continue on my way without any problem.

I'm always amazed by the effect such a small piece of metal can have when put in the right spot.

> Note: This method can relieve cramping for many diabetic patients. Also, if you place a No. 6 silver pellet on F3 on both hands, you can prevent muscle cramps while sleeping.

A Cramping Sole Instantly Relieved

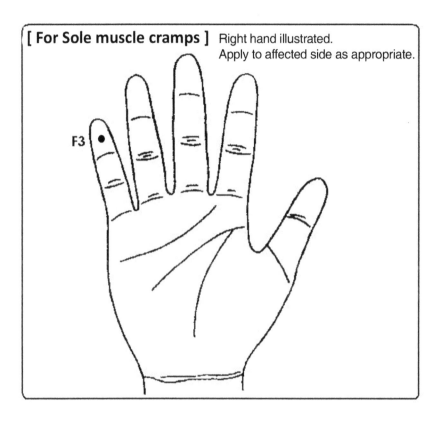

33

―――

Curing an Injured Hip

THE PHONE RANG AT EIGHT o'clock in the morning.
"Who's calling this early in the morning?" I wondered.
A middle-aged woman spoke, "Hello. Is this Dr. Oh?"
"Yes, it is."
"Doctor, my name is Natalie. I'm calling from Toronto. I'd like to make an appointment to see you."
"Can you explain your problem?"
She explained that she was a psychology teacher in a Toronto high school. The week before, she had been carrying a pile of books from the library to her classroom. She didn't realize that there was water on the floor, and she slipped and fell down with both legs spread. Since this fall, her right hip joint had hurt so much that she couldn't walk normally. Sometimes, it was difficult for her to stand for very long. She

went to the hospital. The X-ray showed no broken bone, but the ligaments were damaged and swollen. The doctor offered two choices for treatment: surgery or painkillers. Since that day, she had been taking painkillers, but the hip joint didn't seem to be healing.

"A friend gave me your business card and said you can cure lower back pain very well, so she recommended I come and get treatment from you."

"Okay, but before you come, I have to give you three warnings. First, you will need at least three treatments—one every other day—so you need to be here for at least five days. Second, because of the location of the treatment, you will need to wear brief-style underwear. Third, please don't take painkillers on the day of your treatment. If you agree, when do you want to start?

"Can I get treatments next Monday, Wednesday, and Friday?"

"How about ten o'clock in the morning?"

"Good."

On the appointed day, Natalie arrived and climbed onto the examination table so that I could check her vertebrae and hip joint. Fortunately, the bone was not out of its socket, but every time she moved her leg, she complained of pain. When I checked the lumbar vertebrae at L5, S1, and the coccyx, she felt pain there. She explained that when she fell, she fell hard on this area.

"You hurt not only your hip joint, but you also damaged your lower back. This is a bigger accident than I thought."

Treatment was divided into three steps:

1. Beginning with the lumbar vertebrae, I placed the blue Genesen on L5 and the red Genesen on the tailbone, and I treated her for two minutes. Then, I treated L5, S1, and the coccyx individually for more time. I placed No. 10 magnets on each point.

2. All around the hip joint—above, below, to the left, and to the right—I used the electronic beam, bringing the leads against each place and treating them for thirty seconds each.

3. Using a probe, I searched for the pain-sensitive points on the base of her right pinkie finger. Moving up the finger, I found nine sensitive points. I marked each of the nine points with a pen, and using the electronic beam, I treated the joint. Each treatment took twenty seconds. Then, I placed a No. 1 silver pellet on each point.

On the back of her right hand, I looked for pain-sensitive points along the "I" meridian. I found four points and marked them. Using the electronic beam on either side of the middle metacarpal bone, I treated each point for twenty seconds and then placed No. 1 silver pellets on each point. I placed a No. 6 silver pellet on B1.

On the second day, Natalie told me, "Doctor, it's very effective. The pain has almost disappeared, and the movement in my leg is much smoother. I can walk quite normally now."

"That's great. After today's treatment, you can start learning to walk slowly without a limp."

The second day's treatment followed the first day's three steps, but for the second step, I used Genesen rather than the electronic beam.

After the treatment, I suggested, "Start to practice walking in a shopping center. Walk slowly, making sure your posture is straight. If the pain comes back, stop, but as long as it's not painful, carry on. Don't overdo it, though. Stop after thirty minutes."

"Thank you. I'll see you the day after tomorrow."

Her walk looked much improved.

On the last day, she appeared much healthier.

"Doctor, yesterday I went to the shopping mall and enjoyed it very much. I bought a lot, too!"

"I told you to go to the shopping mall to practice walking, not to buy anything."

"Doctor, do you think a woman could go to a shopping mall without buying anything?"

"Then you practiced walking unconsciously. How is your hip and lower back?"

"I didn't have any problems. I'm perfectly well."

The third treatment was the same as the second treatment. When it was over, Natalie stood and picked up my business card.

"Thank you, Doctor. Goodbye."

"I'm happy to see you cured. Be careful."

I watched her walk away with a perfect stride.

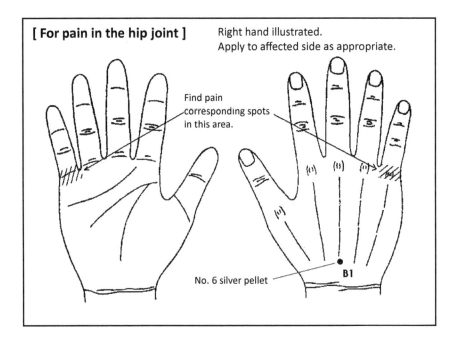

34

Treating a Woman Who Had Fallen from a Tree

One of our Korean friends called and said, "An aboriginal woman asked me to find her a doctor. She injured herself when she fell from a tree. Is it okay if I send her to you?"

"Yes, go ahead."

About thirty minutes later, a taxi pulled up in front of my house. I watched the forty-five-year-old woman walk toward the front door. Her body was leaning to the right about ten degrees, and she was limping.

"I understand that you fell from a tree. How high was it?"

"About one and a half meters (5 ft.)."

"Did you have an urgent reason to climb up in the tree?"

"I was trying to get my cat, but then …"

All of a sudden, embarrassment must have caught up with her because she didn't complete the sentence.

"Where did you get hurt?"

"My right knee and my lower back."

"Please undress and lie facedown on the examining table."

From lumbar L1 through sacrum S1, I pressed each vertebra with overlapped thumbs. She felt pain when I pressed at L3 and L4, so I put little pen marks there. I put the blue Genesen on L3 and the red Genesen on L4, setting both perfectly perpendicular to her spine. I treated the spots for exactly one minute. Then, I put No. 10 magnets on each bone.

I asked her to sit up, and I checked her right knee. I checked in all three directions—top to bottom, right to left, front to back—with strong pressure from my thumb and found a few painful points. I marked the spots with a pen. Then, I aimed the black and red leads of the electronic beam facing each other through the joint so the stream of electrons passed through. I treated each pair for one minute. After treatment, I placed a No. 10 magnet on each of these points.

On her right hand, I looked for the pain-corresponding points using the probe along the length of her "I" meridians[*] on both sides on the middle metacarpal bone. I probed from I15 to I24. I found two points on each meridian and marked them with a pen. Placing the black lead of the electronic beam on one side of the "I" meridian and the red lead to the matching point across the middle metacarpal bone, I treated each pair for twenty seconds. After that, I placed No. 1 silver pellets on each mark.

---

[*] For "I" meridian, see appendix D.

Still on her right hand and using a probe, I looked for pain-sensitive points around the second joint of her pinkie finger. I found pain-sensitive points all around the joint. Placing the black lead and the red lead of the electronic beam facing each other across the joint, I treated each spot for twenty seconds, continuing around the joint. I created a strip of No. 1 silver pellets by putting the metal portion on a strip of surgical tape. Then, I wrapped the strip around her joint like a ring.

I had her stand facing me, about two meters (6 ft.) away. I asked her to put her feet about thirty centimeters (1 ft.) apart and raise both hands over her head.

"Gradually bend from the waist until your hands are on your knees. Do you have any pain?"

"No."

"Now straighten up again and raise both hands to your sides at shoulder height. Slowly turn left about ninety degrees. How's your lower back? Any pain?"

"No."

"Twist to the right. How's that?"

"No pain. Doctor, my back seems to be quite comfortable."

I asked her to stand beside the examining table, holding the edge of the table with her left hand.

"First, raise your right knee, and then swing your bent leg back. Do it again, two or three times."

"Doctor, my knee is okay now!" She looked so happy.

"You need one more treatment. Can you come tomorrow at the same time?"

"Yes, I can. Thank you, Doctor. Sorry to ask a favor, but would you please call me a taxi?"

"Where do you live?"

"Bonnie Doon."

"Then you can take a Number 4 bus at the front of my house. Here's a bus schedule. Look, there'll be one coming in about ten minutes. Why don't you take it?"

When I watched her walking away from my house, her posture was straight, and she was no longer limping.

The next day when she came, she was riding a bicycle.

"Look, Doctor! Now I can ride my bicycle! It's miraculous that I healed so well in only one treatment!"

"You mean you rode your bicycle all the way from Bonnie Doon? That's almost ten kilometers (6 miles)!"

"Doctor, no problem!" she replied with a smile.

This day's treatment was the same as the day before. After the treatment, she rode off on her bicycle just like a young girl.

35

Curing Hemorrhoids

Hemorrhoids can be, in a sense, an occupational hazard. Someone who works in a standing position for a long time, someone who sits for a long time, or someone who is always stressed tends to get this kind of disease. I, who taught for forty years, was no exception. In my third year of teaching, the condition was very benign, but I had hemorrhoids.

At the beginning, I had only a small protuberance, and after passing a stool, there was no problem if it went back inside. If it stayed outside, it was painful, itchy, and uncomfortable. If it stayed outside long, the protuberance might start to bleed. It was very stressful.

Another kind of hemorrhoid affects the internal wall of the rectum, where a polyp develops. Whenever you pass a stool, the polyp could block the passage. This kind of hemorrhoid requires surgical removal of the polyp.

Another kind of hemorrhoid involves the prolapse of the anus. Weak or sickly people and seniors commonly suffer from this variety. If the stool is hard and requires a long time on the toilet to expel, this kind of hemorrhoid is more likely to develop.

Except for the polyp kind, other hemorrhoids can be treated fairly easily by the use of No. 6 silver pellets. I usually use No. 6 silver pellets whenever I feel uncomfortable around the anus, placing the pellets at the bottom of the middle metacarpal bone near the wrist. You can easily find the little dip at the end of that bone. That point is B1. Place pellets on both hands and leave them there for three days. In doing so, all the symptoms disappear very quickly. If you place them there at bedtime, the next morning, you'll feel very comfortable. Long ago, I threw out the suppository ointment that I had been using. Hemorrhoids no longer stress me.

Two factors aggravate hemorrhoids: (1) constipation and (2) sitting on the toilet for a long time.

Enjoy your life without hemorrhoids.

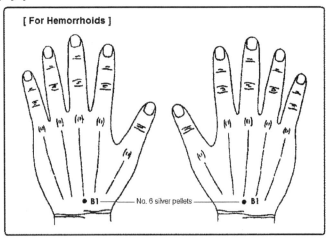

36

Curing a Restless Child

THE CITY OF FORT MCMURRAY, about 450 kilometers (280 miles) north of Edmonton, is well-known for its oil industry. Heavy oil and natural gas are abundant there.

A Caucasian woman from the city called to ask, "Doctor, would you please look at my son?"

"What's the problem?"

"There's no pain or anything; the only problem is that he can never sit still and is always causing problems."

"How old is he? Most young boys act like that, don't they?"

"My boy is five years old, but he has always been restless. He never obeys the teacher and is very wild in his play. He's been kicked out of kindergarten and seems to be getting worse recently. My mother, who lives in Edmonton, told me about you. You cure many difficult illnesses, so she suggested that I bring my boy to you. Would you please take a look at him?"

"When can you come?"

"It takes about six hours to drive to Edmonton, so I can come around three o'clock in the afternoon."

"Good. Can you come around four o'clock tomorrow? Let's make an appointment for then."

The next day at around 4:00 PM, a woman in her thirties arrived at my door with a boy.

"Doctor, meet Kevin."

Kevin looked very clever. While his mother took a few minutes to fill out the diagnostic form with his name, address, and date of birth, Kevin left the office and went through all of the rooms of the house, even the basement.

According to my examination and his pulse, his liver was overactive and his gallbladder was deactivated. I used the electronic beam to deactivate his liver meridian. In other words, I placed the black lead on N9 of his right pinkie finger and the red lead on N5 and treated him for twenty seconds. I treated both hands the same way. While I was using the electronic beam, Kevin was very quiet and attentive, obviously curious about the machine that was emitting a strange beeping noise.

I asked his mother, "Do you drink tap water?

"Yes."

"No matter how well the waterworks system cleans it, tap water always has pollutants mixed in, particularly when there are such big chemical factories nearby like in Fort McMurray. The pollutants are not only in the air, but in the water, too. It is impossible to filter out all of the pollutants from the water. This is my hunch, that Kevin's body is polluted by heavy metal. The body tries to detoxify itself through the liver of all the toxic materials that enter it, but if pollutants keep entering, the liver cannot manage all of them. Not only that, but the liver is weakened and can create hepatitis or something similar."

"Then what can we do?"

"The best way is to leave the city, but if that is impossible, you have to buy bottled water. I understand that there is a water company that will deliver a five-gallon bottle to your home. Order distilled water from which all of the pollutants have been removed, and you'll be drinking clean water."

The next afternoon, I got a phone call from the woman.

"Doctor, thank you for yesterday. After that treatment, Kevin changed completely. On the way home, he was so quiet that I thought I was the only one in the car!"

"That's great to hear. Thank you for reporting this to me."

"But, Doctor, can I come back the day after tomorrow at four o' clock? I'd like to see you again. Is that time okay?"

"Yes, you can make an appointment, but Kevin is okay, right? Why do you need to see me again?"

"I have two more boys: a seven-year-old and a three-year-old. I'd like you to see them."

She brought the two boys on the day of the appointment. I checked their pulses. The older boy had the same condition as Kevin: his liver was overactive and his gallbladder was deactivated. According to his mother, the older boy acted almost the same as Kevin but was not making too much trouble at school. I gave him the same treatment as I did Kevin.

The younger boy had an almost normal pulse.

"This boy is okay. There's nothing wrong with him."

"Wow, that's fortunate. This boy has never acted like his older brothers, but I brought him anyway because I was worried about him."

This shows how much influence pollutants have in children's bodies. Treatment is very easy.

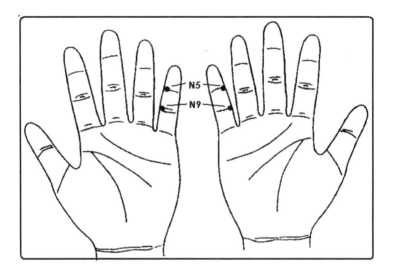

37

A Student Who Miraculously Escaped Renal Failure

My niece in Tokyo called. She's a third-year medical student, and her best friend in class had suddenly been admitted to a hospital.

My niece visited her friend and learned that due to renal failure, her friend couldn't pass urine. Her face had darkened, and her body had started to swell. According to the doctors, she had a very acute problem. Even though she got all kinds of treatment and drugs, she still couldn't pass urine. Her doctor was quite confused.

My niece asked if I knew how to take care of such a problem.

"Do you remember when I was in Tokyo last time and gave you a set of therapeutic rings to control menstruation? Do you still have them?"

"Yes."

"I want you to put your rings on your friend's pinkie fingers. Because she's under a doctor's care, you'll need to get permission from her doctor before you do that."

The next day, she called again.

"I asked the doctor's permission to put the rings on. The doctor wasn't too enthused and said, 'What can those little rings do? Go ahead.' He said, 'It doesn't matter anyway. There won't be any effect.' So I put the rings on my friend's fingers. The next morning, she passed quite a bit of urine, and the doctor exclaimed, 'How can a little ring have such an effect? I can't understand the function of this ring.' My friend was discharged that day."

When the rest of my niece's medical class heard this story, many students wanted to buy therapeutic rings, so I had to send sixteen rings by special delivery.

If her friend hadn't passed urine that day, she would have had to start dialysis, and according to the doctor, once you start dialysis, you have to do it for the rest of your life.

38

Curing a Young Woman's Pimples

ANN WAS AN ELEVENTH-GRADE STUDENT who dreamed of becoming a model. She heard that a representative from an American modeling school was coming to Edmonton to recruit students, and she decided to apply. She was worried, however, about the pimples on her face. She had tried many different treatments, but they had little effect. She even saw her family doctor and tried some prescription cream, but it didn't do any good either. Her mother decided to bring her to me.

Ann looked more mature than her seventeen years. I noticed about twenty pimples on her face, mostly on her forehead and cheeks.

"So when is your interview and photo session?"

"Exactly one month from now," Ann answered, but her mother immediately asked, "Can you get rid of them?"

"Oh, yes. Some people get a quick effect, and some take longer, but you have one month, so I don't think you'll have a problem."

According to her pulse on her right, she had an overactive liver, and on the left, her stomach was overactive. The treatment on her right hand required No. 1 pellets to be placed according to the *liver-excessive* formula plus E38. On her left hand, I placed the pellets according to the *stomach-excessive* formula plus E38, G13, and N3. I told her to keep the pellets in place for at least forty-eight hours if not three or four days. I gave her two bottles of Chinese medicine—*bonyujang*—a capsule containing 200 mg of royal jelly and 15 mg of ginseng.

"Take one capsule in the morning and one in the evening. Because there are thirty capsules in a bottle, two bottles will last exactly one month. If your pimples don't disappear within two weeks, come visit me once again." That finished the treatment.

About a month later, her mother called saying, "Doctor, the pimples on Ann's face disappeared completely, and she passed the entrance exam and photo shoot. Thank you very much."

Curing a Young Woman's Pimples

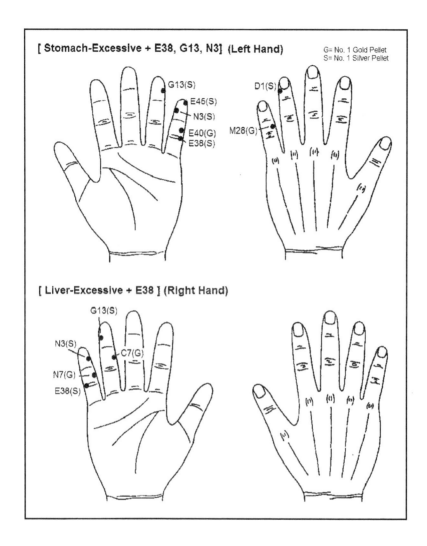

39

A Woman Suffering with Menstrual Pain

I GOT A PHONE CALL from a young immigrant woman from Colombia. She wanted to have me make a house call to see her two-year-old son, who had severe asthma.

After the treatment, I packed my bag and was ready to leave when she asked, "Do you have any way to treat menstrual pain? I suffer such pain that I have to stay home from work."

"That's relatively easy."

I gave her therapeutic rings for each pinkie finger.

"Put the rings on about a week before you expect your period to start. Then, leave them on until the end of your period. You probably won't feel any pain. Even if you forget and then put the rings on once the period starts, you'll still get a good effect."

About a month later, she called and said, "Doctor, those

rings are miraculous. Even though I had my period, I didn't feel a thing. Thank you for teaching me such effective pain control. I'm sharing the rings with my mother and sister. All three of us are using them now."

"That's great, but with only one set of rings, I hope that all three of you don't get your periods at the same time!" I laughed.

40

Curing a Patient Who Didn't Believe in Alternative Medicine

IN EASTERN CANADA, THERE'S A province called Newfoundland and Labrador. St John's is the capital, and a Korean professor named Dr. Park at the university there used to be at the university in Edmonton, where we originally met.

I got a special opportunity to visit St John's. From Edmonton to St John's, the time difference is three and a half hours, the distance is 6400 kilometers (4000 miles), and the flight takes eleven hours; it's not easy to visit.

When my wife and I visited Dr. Park, he and his wife welcomed us, but I noticed that Dr. Park's right eye was badly swollen. His eyelid was almost half closed.

"Dr. Park, what happened to your eye?"

"I didn't get an injury; it must be bacteria. It's been tearing up a lot in the past few days, but today I started to get some pus in the eye. I went to a doctor this morning and got some ointment."

"How long has it been since you put in the ointment?"

"Three hours."

"Are you feeling any better?"

"Not much."

"I'm practicing Korean hand therapy. Let me cure your problem."

"Oh, I don't believe in that kind of stuff."

"The treatment works regardless of whether the patient believes, so let me cure you."

"Doctor, I'll never believe that, so it's useless. Don't bother."

"Dr. Park, I'm not using a knife or needle or any instrument. I'm not even going to touch your eye. I'm only treating your hand. You have nothing to worry about. Just give me your right hand."

"Really? You're not going to use acupuncture? Okay, then please go ahead."

Finally, he gave me his right hand. Using a probe, I touched the tip of his middle finger at E2. There are two E2s in that finger, and I chose the E2 on the side toward his pinkie finger and pressed a bit with my probe.

"Do you feel pain here?"

"Yes."

"Give me your left hand."

This time I used a probe on the E2 closer to his thumb and pressed a bit.

"Do you feel pain here, too?"

"Yes, I feel it there, too."

I brought out my electronic beam and placed the leads on the two painful E2s—one on each hand—giving them thirty seconds of treatment. Then I placed No. 1 silver pellets on each point.

Because we had not been together for a long time, we enjoyed dinner together that evening. We promised to go sightseeing the next morning, and my wife and I returned to our hotel.

The next morning when I saw Dr. Park, I noticed that his right eye looked normal. It was neither swollen nor red.

"Look, your eye's already cured."

"Doctor, your treatment is really remarkable."

"Whether or not you believed, this worked very effectively. Someone once said, 'Whether you believe it or not, the earth turns.' I'm glad your eye is now cured."

"Now I can believe in your treatment."

After four days, the meeting that had taken twenty years to come ended, and we had to say our goodbyes again.

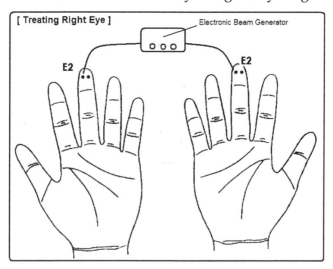

41

Discovering How to Stop a Nosebleed Instantly

On a visit with my brother in Los Angeles, we were enjoying family conversation after dinner. All of a sudden, my nose started to bleed. The amount of blood was so unusual; there was a lot of blood. I tried to block it with a small cotton ball, but that didn't work. I tried to pinch my nose between my thumb and index finger, but that didn't work. I wrapped an ice cube in a cloth and put that on my nose, but that didn't help. I asked someone to hit the back of my neck quite hard, but that didn't work.

I remembered that I'd read in a book a suggestion to tie a rubber band around the first joint of the middle finger, making it so tight that the tip of the finger becomes almost black. I tried that, but the pain at the fingertip was unbearable. I waited another five or ten seconds until it was impossible to bear and quickly removed the rubber band. According to the book, the nosebleed should have stopped at that moment, but it didn't.

Then my wife gave me a 7.5 cm (3 in.) square of aluminum foil and asked me to wrap it around my middle finger. I wrapped my left middle finger first, and the blood coming through my left nostril stopped instantly. My wife gave me another piece of aluminum foil, and I quickly wrapped my right middle finger. That stopped the bleeding from the right nostril.

We were all thankful that such a scary nosebleed had stopped. We stayed quiet. I lay down on the bed for about twenty minutes, wondering how a piece of aluminum foil could stop a nosebleed. This was a new discovery and very wonderful.

I sat up on the bed. It had been about forty minutes since the nosebleed started and twenty minutes since it had stopped. Because it was summer, the inside of the aluminum foil was sweaty and uncomfortable. I was wondering how long I should keep the aluminum foil on. What would happen when I removed it? I wondered whether the bleeding would start again. Normally, when you have a small cut or injury, it takes about ten minutes for blood to coagulate and stop the bleeding, so applying the same theory, twenty minutes should be enough for a nosebleed.

Still, I decided to wait thirty minutes. Then, I held the tip of the aluminum foil on my right finger and pulled. Instantly, the bleeding restarted in my right nostril. Almost frantically,

I put my finger back into the cylinder of aluminum foil. The bleeding stopped again.

I waited another half hour and tried again. I removed the aluminum foil from my left finger first. It didn't bleed, so I pulled off the other aluminum foil. This time, I realized that the bleeding had completely stopped.

With this experience, I discovered that the left middle finger controls the left nostril and the right middle finger controls the right nostril. I have tried this method many times since this incident. Now I'm certain that it works every time.

Since then, I always carry 10 cm (4 in.) squares of aluminum foil in my pocket diary.

42

Treating Eczema on the Scalp

Susan was a twenty-three-year-old university student. She came along with her mother and asked for my help. Her mother said, "Doctor, my daughter has been suffering from eczema on her scalp. It is so itchy that she can't concentrate on her studies."

"Susan, have you seen a doctor? What did the doctor say?"

"Yes, a dermatologist confirmed that it's eczema. She prescribed some ointment. When I put on the ointment, the itchiness subsides some, but then a scab forms, and it gets so itchy that I can't help scratching. Once I touch it, there's a watery secretion, so it's very difficult to leave it alone. I can't study, and I can't sleep well."

When I looked at her scalp, I saw that the eczema had spread all over her head, and I noticed cracks here and there. I saw some very wet areas caused by the secretions. Fortunately, the eczema covered only the area with hair and didn't spread to her neck or face.

I placed pellets on Susan's hands according to the *lung-wet* formula.

"Within two or three days, you'll see some difference, but you need to keep the pellets in place for at least four days."

A week later, Susan's mother called me to say, "Doctor, Susan's eczema is completely healed. Thank you very much. I wish we'd known you earlier. She's suffered so long." She sounded very happy.

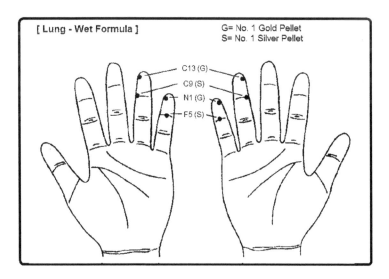

43

Curing Sudden Ear Pain

IN VANCOUVER, WHICH IS KNOWN as one of the most beautiful harbors in the world, we had a church meeting with many clergy and lay representatives in attendance.

On the first day, we gathered in a big circle in a classroom. The president of the meeting gave an opening statement. After about five minutes, the wife of the Rev. Kim, sitting directly across the circle from me, suddenly put her hand to her left ear, massaging it. Her facial expression showed sudden pain. Because she was sitting at the side farthest from the classroom door, if she wanted to leave the room, she would have to cross the circle. Perhaps she didn't have enough courage to cross the room during the president's address. She was wiping away tears with a handkerchief, but still she didn't move.

Fortunately, the president soon finished his opening statement, and before we got into our continuing education program, we had a fifteen-minute recess. Mrs. Kim left the room very quickly. While I was visiting with old friends over coffee, I noticed Mrs. Kim reenter the room. I went over and asked, "Are you having trouble with your ears?"

"Yes, all of a sudden my ear was very sore. I didn't know what to do."

"Have you had the same problem before?"

"No, this is the first time."

"Mrs. Kim, may I help you?"

"Yes, please!"

I asked her to extend her left hand, and then, using my probe, I tried to find the pain-sensitive corresponding point in the middle of the tip of her left middle finger, gradually moving across the pad toward her pinkie finger. Suddenly, she pulled her hand away. "Ouch, that's painful there."

I placed a No. 1 silver pellet at the point.

I did the same thing on her right middle finger, this time moving toward the thumb side of that finger pad. I pressed that point with my probe, and she said, "It's painful there, too," so I placed a No. 1 silver pellet there.

"How is your ear now? Does it still hurt?"

"Wow! It's not hurting any more. I don't know how that happened, but it's healed. It's amazing."

"Keep those pellets in place for three days. Then you can take them off."

"Thank you very much for curing my problem so quickly. That was really amazing."

During lunchtime, the Rev. Kim came to see me, asking to shake my hand. "Thank you very much for helping my wife."

44

Curing Acute Tonsillitis

THE REV. BAI IS A minister in a Korean church in Sao Paulo, Brazil. Because they speak Portuguese in Brazil, my wife and I had a lot of trouble communicating. The Rev. Bai helped us while we were there.

One Saturday, I invited the Reverend and Mrs. Bai for dinner. At the dinner table, he commented, "I have a meeting at eight o'clock tonight. I was supposed to give a speech at the meeting, but I have a problem with my throat."

I had already noticed that his voice sounded different. "Does your throat hurt?"

"Yes, my throat is hurting, and it's difficult to talk."

I looked into his throat. His tonsils were swollen, and the whole area was red.

I placed a No. 1 silver pellet on both of his hands at A20, A22, A23, A24, and B24.

"You still have more than an hour, so you have nothing to worry about. You'll be okay by then." After that, we went back to our hotel.

The next morning, my wife and I attended Sunday service and heard the Rev. Bai preach. I noticed that his voice had returned to normal, and I asked, "Reverend, how was the speech last night? Your voice sounds normal now. It must be all healed."

"When I was giving the speech, I noticed that my voice had returned to normal already. By the way, many people speak of instant healing, but really, this is so quick. I never realized there could be such a quick remedy. This is really amazing." His voice rang clear.

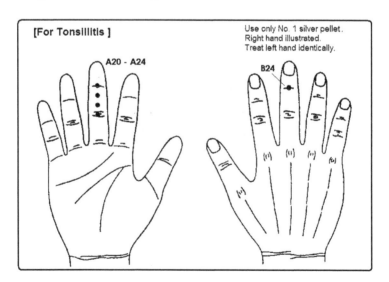

45

Realigning a Jaw That Was Causing Molar Pain

Barbara has been my patient for some time due to chronic lumbago. One day, she called and said, "Doctor, how are you? It's Barbara. I'm calling for a favor."

"Hello. It's been a long time. Any problems?"

"One of my molars hurts."

"I'm not a dentist!"

I was ready to hang up the phone, but she said quickly, "Just a minute! When I went to the dentist, he told me that my jaw was out of alignment. That's why my molar hurts, so he suggested I get my jaw joint adjusted."

"Oh, in that case, can you come tomorrow morning at ten o'clock?"

Barbara showed up on time. She looked much healthier than before.

"Doctor, look at my left jaw. It's out of alignment."

I asked her to sit facing away from me. I put my hands on both jaw joints and asked her to open her mouth as wide as possible. The left jaw felt like it moved forward a bit.

Treatment was simple. Turning her around, I asked her to extend both hands, palms up. Starting at the center of the pad of her left middle finger and moving toward her pinkie, I used the probe to find pain-sensitive points on that finger. Almost at the side of her finger, she complained of pain, so I placed a small mark there. On her right middle finger, I moved from the center of the finger pad toward her thumb. On the thumb side of that finger, she again complained of pain. I put another mark there. Placing the black lead of the electronic beam at the mark on the left finger and the red lead on the mark on the right finger, I treated her for forty seconds.

I placed No. 1 silver pellets on each point. I asked her to turn her hands palms down, and I placed No. 1 silver pellets on B24, the first joint of the middle finger on both hands.

She turned, facing away from me again, and I put my hands on both sides of her jaw as before. I asked her to open her mouth and noticed that the left jaw stayed properly in place even when her mouth was wide. She repeated the movement several times, and the jaw never went out of alignment.

Barbara was amazed at the quick remedy.

"Doctor, I'm certain you're a miracle worker."

I was glad to see that she was leaving so happily.

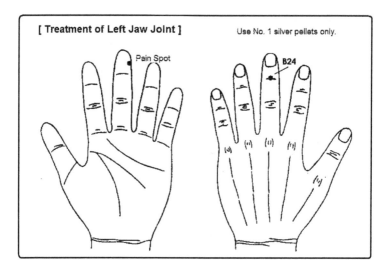

46

Curing a Hockey Player's Dislocated Spine

A YOUNG MAN CAME INTO my office and said, "You must be Dr. Oh. I'm Peter. Please help me." His voice was rather weak.

"Where do you have a problem?"

"I'm a hockey player. At my last hockey game, one of our opponents bodychecked me against the boards so hard that it hurt my back. Since then, I've been having problems breathing, and I've lost all my strength. Doctor, would you help me out?"

"Well, let me see. Take your shirt off and sit here."

I touched his back from the shoulder blades down. I could feel that four vertebrae were pushed in. I asked him to lie facedown on the examining table, and I checked him carefully. His thoracic vertebrae at T9, T10, T11, and T12 were pushed forward. I diagnosed dislocation of the spine.

"Peter, four of your vertebrae have been pushed in. I think I can help you. Do you want me to start right away?"

"Yes, please."

Treatment involved placing the blue Genesen at T8 and the red Genesen at L1 and treating for about five minutes. Then, I treated the individual vertebrae for one minute each with Genesen. After that, I placed one No. 10 magnet on each vertebra from T8 to L1. On his hands, I used the probe to look for the pain-sensitive points on his "I" meridian* from near I14 through I24. I marked three painful points on each "I" meridian for a total of six on each hand.

Starting with the black lead of the electronic beam on the top spot on one meridian and placing the red lead on the matching spot on the other side of the middle metacarpal bone, I treated each pair for forty seconds. When I finished the electronic beam treatment, I placed a No. 1 silver pellet on each point.

"The treatment is over, Peter. We have to wait now to see if you get a good effect. Give me a call when you feel some change."

"Yes, I'll do that."

"You don't have to worry too much. It will come back. Go easy, though. Don't play hockey yet, but believe that the problem will be cured."

"Yes, I understand. When shall I come to see you again?"

"How about in a week, at the same time? I'll make an appointment for then, if that's okay."

"Yes, fine. I'll see you then."

---

* For "I" meridian, see appendix D.

The young man left.

The next day, Peter called. "Doctor, I heard a click-click in my back last night. Is this my bones going back into place?"

"Peter, you're a lucky man. You're already getting the benefit of treatment."

"Doctor, is this a good sign?"

"Yes, it is. Give me a report again tomorrow."

Actually, I was surprised at the quickness of the effect.

The next day, Peter called back and said, "Doctor, I feel normal again. I asked one of my friends to check my spine, and he says my vertebrae are all straight. My energy seems normal again, too. Do you think I'm all better?"

"Peter, I'm amazed at how quickly you've returned to normal. If you get another impact on that same spot, though, your vertebrae will come out of alignment very easily, so you have to be very careful about taking another blow to that same spot. As long as you remember that, you'll be okay. Let's cancel next week's appointment."

"Good! Doctor, thank you very much. I won't forget you for the rest of my life. Goodbye."

Peter sounded very happy.

47

Curing Pancreatitis in a Simple Treatment

On a visit to Seoul, my wife and I were invited out by a friend, Mr. Kim, president of a trading company. "Do you have any free time this evening?" he asked.

"Yes, today's Sunday, so I have lots of time."

"Let's have dinner together, and after dinner, can you spend some more time?"

"That's fine. What do you have in mind?"

While we were having dinner, he told the following story:

"My brother-in-law has acute pancreatitis. He's been admitted to a hospital and is scheduled for surgery tomorrow. I want you to look at him tonight, if possible. He is a well-known university professor who carries important responsibilities in the arts. If he has to stay in the hospital for more than a month, the schedule for public performances will be badly affected."

After dinner, we went to visit his brother-in-law at the hospital. The very famous Professor Park was lying in bed with his daughter at his bedside. After our greetings, I asked, "Does it hurt very much?"

"No, it doesn't hurt much, but I feel very heavy."

"Let me check your pulse."

With this illness, one can't touch the abdomen, so I just checked his pulse. On both sides, the pulse showed that his gallbladder was overactive and his liver was deactivated. "Professor, your pulse says there is an issue with more than your pancreas: you have a gallbladder problem."

"Yes, I have a gallstone. What can I do?"

"I can treat you to calm your pancreas. This treatment is on your fingers, not your body. If this works, it will show up in tomorrow's presurgery test. Do you want me to treat you?"

Without hesitation, he said yes.

The treatment was quite simple. Keeping the area and my equipment clean with alcohol, I drew four or five drops of blood from the end of each pinkie at F1.

"Now that that's over, I'm all finished. Have a good night's sleep. If anything happens, please call."

We went back to the hotel.

The next day, Professor Park called.

"Well, what happened? Where are you calling from?"

"I'm calling you from home. I was discharged this morning."

Before I could say anything, Professor Park continued, "The hospital said I didn't need the surgery, so they sent me home. By the way, do you have time this evening? Why don't you come to my house for dinner?"

"Yes, I will."

That evening at his house, Professor Park and his wife prepared a big meal. They both thanked me very much for such a quick and easy fix to a difficult illness.

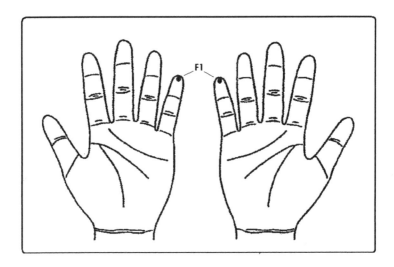

48

Curing a Lump on the Chest

Sarah, an Italian immigrant who has been living in Edmonton for about twenty years, practices law in the city. Two years ago, she was in a car accident and injured her lower back. About two months ago, she was in another car accident and injured her neck and right shoulder. She wanted me to cure her lower back, neck, and right shoulder.

The lower back and neck were cured very quickly through my specially developed and patented method, but her right shoulder pain didn't disappear even with my effective methods. I didn't understand. I asked Sarah to take off her blouse and found just below her right collarbone a 3 cm (1.25 in.) lump, about the size and shape of a small egg. It was protruding from her chest.

"When did you get this lump?"

Sarah was surprised at my question. "I didn't even know it was there!"

"You didn't notice when you took a shower?"

"No, never."

"No wonder your shoulder pain isn't subsiding. Let's cure this problem first. Do you want me to start treating this lump right away?"

"Yes, please," she agreed eagerly.

I put the black lead of the electronic beam at the lower side of the lump and the red lead on the upper side and treated it for thirty seconds. Placing the leads of the electronic beam on both sides of each E12* on each hand, I treated those spots for twenty seconds. Then, I placed No. 1 silver pellets on both of her middle fingers at both E12s, four silver pellets altogether.

I made an appointment to see her a week later.

Two days later in the morning, Sarah called and said, "Doctor, I'd like to see you today. Can I come right now?"

"What's the problem?"

"I can't talk on the phone. I have to meet you."

"You have an appointment for next week. Why don't you come then? I'm busy with another patient now."

"Doctor, I can't wait that long, I have to see you today."

Because she was so eager, I thought that there must be a big problem. I looked at my appointment book.

"You can come to see me at 3:30 this afternoon. Is that okay?"

"Yes, I'll be there."

Sarah was there right on time.

---

* There are two E12s on each hand.

"What's the problem?"

She didn't answer my question, just pulled her blouse away from her collarbone. "Doctor, look. It disappeared."

"Wow, that's wonderful. Congratulations. But is that why you wanted to come and see me in person?"

"Where did it go?" Sarah asked very seriously. "It might be hiding somewhere and come out later in a different place. Would it?" She looked worried.

"No worry. It must have dissolved and been absorbed by your body. It might have passed away with your urine. You have nothing to worry about. By the way, how is your shoulder pain?"

"My shoulder pain has disappeared."

"Sounds like we've got a miracle. I'll cancel next week's appointment, okay?"

"Yes, please. I never expected all of these problems to disappear so quickly." She was so happy.

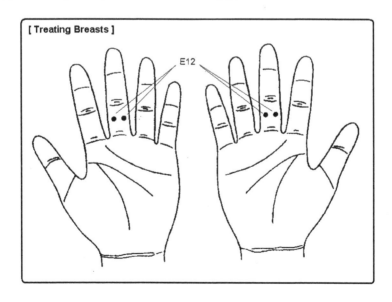

49

A Seasick Young Man

Leaving from Miami in the southeast United States, we were heading to Brazil on a cruise ship. One day at lunch, my wife and I went to a nice restaurant and filled our trays with food from the buffet. Because we had arrived at the busiest time, it was hard to find a table open. Then we noticed in a far corner a big round table with only one occupant. I asked the young man if we could join him, and he very pleasantly agreed.

Usually in such cases, we exchange pleasantries, asking questions like, "Where are you from?" "What do you do?" and "What nationality are you?"

Observing the man eating his lunch, he acted as if he were eating something bad or that someone was forcing him to eat it. I wasn't very patient and asked, "Do you have a problem? I don't like the way you eat!"

He scowled, "Yes, I'm having a bad case of seasickness. I've been suffering for a week, and I'm almost dying. We still have ten days to go, and I don't know what to do."

"Is that right? I practice alternative medicine and can easily cure that problem. If you don't mind, after lunch, I'll get my treatment equipment, and we can meet somewhere. You pick the place."

"You can do your treatment anywhere?"

"Yes, I'll put small pellets on your hands, and you'll be okay very quickly."

"In that case, can we meet at 1:30 PM on the left side of the pool? There are lots of chairs there."

At 1:30 PM, he was already waiting for me when I arrived at the pool. I placed No. 1 silver pellets on both hands at A8, A12, A16, K9, and F4. While I placed the pellets, he watched with great interest. When I finished, he asked, "How in the world can you fix an illness with such small pellets?"

"All of our internal organs and parts of our body are connected to 345 points in each hand. If you fix the points in your hands, the connected organs will be normalized." Saying that, I gave him my business card, and he gave me his. I said, "As you can see on the back of the card, I specialize in many difficult illnesses."

"You can fix all of these illnesses?"

"Usually in one or two treatments, they are completely corrected."

"By the way," he said, "my dizziness disappeared. My stomach feels very comfortable. It's amazing.

"I see the word Korean here. Are you Korean?"

"Yes, I'm Korean, but we live in Canada."

"I'm Brazilian and live in Rio de Janeiro. I'm glad to meet a Korean."

Hearing mention of Rio de Janeiro, I said, "We wanted to visit the site of Christ the Redeemer on top of the mountain but don't know who to ask about how to get there and how long it will take. We were looking for anyone who could tell us about sightseeing in Rio."

"Don't worry about sightseeing in Rio. Look at my business card. I'm the vice president of a big jewelry company and can have one of my English-speaking staff guide you with our company car to all the sites. But I have a question to ask about Korea. Can you help me?"

"What's that?"

"There's an international jewelry exhibit in Kyungju, Korea. We'd like to show our product there, so we have many questions about the country. If you could help me, I would really appreciate it."

"You can ask me anything. We still have ten days on this boat, so you can ask any questions that come to you."

Unexpectedly, we had a new friend. Our sightseeing in Rio was extremely interesting and exciting.

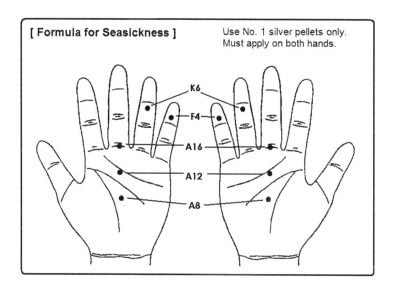

50

Curing Severe Seasickness

On a one-week cruise in the eastern Mediterranean, we were assigned a dinner table for four with another couple. On our first evening, we introduced ourselves and talked of how we could get the most out of the week. The other couple was from Cleveland; the husband was Caucasian and the wife, Filipina. We enjoyed our first dinner together and went our separate ways.

The next evening when we arrived at the table, the husband was there alone. We assumed the wife was busy and would join us later. While we were talking, the waiter came and handed out the menus. He put a menu down at the empty place, but the husband handed it back, saying, "My wife can't come for dinner tonight." I asked him if something had happened. He explained, "She's very seasick. She's thrown up a lot and is so dizzy that she can't leave the cabin."

"You mean she hasn't eaten breakfast or lunch?"

"That's right. We've planned this cruise for such a long time and were looking forward to an enjoyable trip. But now, on the first day, she's so seasick. I don't know what to do. All of our wonderful plans are spoiled." He looked very discouraged.

"I practice alternative medicine, and seasickness is a simple illness to cure. After dinner, why don't you choose a place where we can meet? I'll treat her hands, not anything else. Tell her to wash her hands and meet me. It will take no more than five minutes."

He thought for awhile. "There's a small table and chairs next to the library. Why don't we meet there after dinner?"

"Good. I'll go back to my cabin to get my equipment and meet you there."

We had an enjoyable dinner and went back to our cabins. He called to say that his wife would be outside the library in ten minutes.

"Okay, we'll meet you there."

When I got to the meeting place, both of them were waiting. She looked miserable.

"Just extend your hands, palm side up." I brought out No. 1 silver pellets and placed them on A8, A12, A16, K9, and F4 on both of her hands.

Then I told her, "Keep the pellets in place until tomorrow morning, and you'll be okay."

All of a sudden, the woman burped and started to smile.

"Wow, my stomach is getting comfortable, and the dizziness seems to be lessening. Thank you. We'll see you tomorrow at the dinner table!"

For the rest of the cruise, she never missed dinner and never got sick again. We had a very pleasant cruise.

When the ship docked at Mykonos, the night scene was like a dream. The four of us went ashore and enjoyed strolling around the island. We found a small restaurant on the beach where we shared beer and fried calamari for our last dinner together.

We said our farewells. All of us hoped to meet again somewhere.

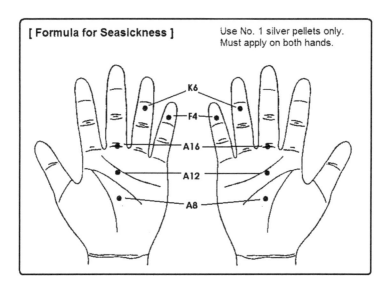

51

―

Curing the Cold of a Visiting Overseas Speaker

Dr. Chang is a well-known scholar in Korean education. She studied at the University of Alberta while I was teaching there. Many years later, she was invited by the Canadian university to deliver a lecture.

When she came to Edmonton, we invited her to have dinner with us the night before her presentation. We talked about the Korean education system and its recent rapid developments.

Dr. Chang said, "Doctor, I have a problem. I'm catching a cold. I feel cold and my throat is giving me problems. I have to deliver my speech at ten o'clock tomorrow morning. I don't know what to do. Can you help?"

"Actually, I noticed your voice changing while we spoke.

For one thing, your clothes are too light for this climate. Did you forget about Canadian weather?"

I brought a thick jacket from my closet and gave it to her. "This is brand new, never worn. Take this with you to keep warm."

Then I took her to my office and put pellets on both of her hands according to a formula that I knew to be very effective for colds. On both of her hands, I placed pellets at H2, I38, A8, A12, A16, A20, A22, A24, A28, B17, B19, B22, B24, and B26.

"It may be cumbersome to have so many pellets on your hands, but keep them there as long as you can, ideally for three days. Also, sleep warmly tonight. Sweating will actually help. You'll notice in the morning that your cold is gone."

After the treatment, I drove her back to her hotel.

The next afternoon, Dr. Chang called and said, "Dr. Oh, thank you very much for treating me last night. This morning, my voice had come back before I had to deliver the speech, and I felt no cold symptoms. My cold was completely cured. I had a very good session and a very pleasant day in Edmonton. I'm really amazed that your treatment worked so well.

"I have to leave early tomorrow, so I can't see you again, but I left the coat you loaned me with a friend who promised to take it to you."

"I'm very happy to hear that your cold is cured and your speech went well. Have a pleasant trip home," I replied.

I was happy to hear that she was able to deliver her lecture. If you were invited from overseas to deliver a special speech, it would be very embarrassing if you couldn't do it due to illness.

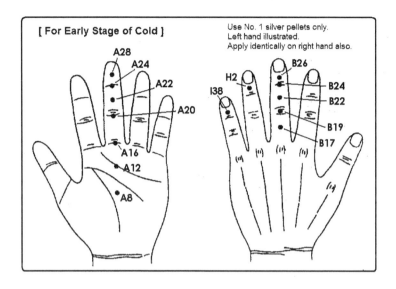

52

Curing a Chronic Headache

THIS HAPPENED WHILE WE WERE on the cruise to Brazil, right after I treated the seasick young man. Many people around his cabin became aware of the treatment.

A sixty-three-year-old Colombian woman came to see me shortly after.

"Dr. Oh, please help me."

"Where do you have a problem?"

"The back of my head has been hurting for a long time." She pointed behind her right ear and at the middle of her neck. I touched the spot.

"You said it's been hurting a long time. How long are we talking?"

"Over two years. I got painkillers from the doctor, but they don't last long. After awhile, it hurts again."

"Are you having pain now?"

"Yes."

"Give me your right hand, palm down."

I placed her right hand on my left hand, and using the probe, I searched for a pain-sensitive point below the nail on her middle finger. She complained of pain on the M5 on the pinkie-finger side of her middle finger. (There are two M5s on that finger.) I tried the same point on her left hand and found that point a little less painful than on the right. I placed a No. 1 silver pellet on the right M5 on her right hand.

After treatment I asked, "How's your headache now?"

"Wow! The pain disappeared." She looked surprised and massaged the spot with her hand.

"Keep the pellet there for three days, and if it hurts again, come and see me."

"Thank you!" she said as she left.

After eight days, when we arrived at our final destination, we lined up to leave the ship. She came to me in line.

"Doctor, I haven't had any pain since your treatment. Thank you very much." She was smiling broadly.

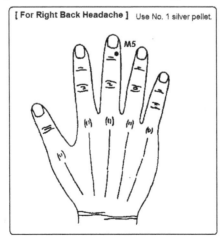

53

Curing Chronic Esophagitis

My wife and I were invited to my brother's son's wedding in Los Angeles. At the reception, my brother-in-law (the husband of my younger sister) sat beside me.

He surprised me by saying, "I'm afraid every time I see food nowadays."

"What? You're afraid just looking at food? Why?"

"For the past three months, I've had esophagitis. I've been taking drugs the whole time, but I can't get rid of it, and every time I eat, it's very difficult to swallow. It even hurts as the food goes down. Is there any way you can fix this for me?"

"Was the medicine prescribed by a doctor?"

"Of course, but I'm really scared: one of my friends has similar symptoms, and his doctor just told him he has cancer."

"Because I brought my treatment bag with me, why don't I treat you right now?"

I brought out the electronic beam from my bag and placed the red lead at A12 and the black lead at A24, and I treated him for thirty seconds. Then, I placed No. 1 silver pellets from A12 through A24, forming one straight line. I did exactly the same treatment on the other hand.

I explained, "The main cause of esophagitis is bad habits. When we get hot coffee or hot soup in our mouth, we just swallow it automatically. Then, the hot liquid actually burns the esophagus. Sometimes, we eat very spicy or salty food. The surface of the esophagus is covered with a sticky film to protect it, but that stickiness dissolves quickly in hot or salty liquid, like soup. Then, the wall becomes bare, and one can easily get an infection from spicy food. Therefore, you need to reduce your intake of hot, salty, and spicy foods as much as possible until the esophagus wall recovers its sticky film. If you continue to eat hot, salty, spicy foods, your esophagus wall will never recover, and in time, it could get cancerous."

About five minutes later, our food was served. Helping himself, my brother-in-law noticed, "This food is going down very easily now. I don't feel any pain around there. It's amazing!"

"Don't relax yet. Remember, you aren't able to put ointment on the wall of your esophagus. You can only control your food intake. Take it easy, and keep the pellets on your hands for at least three days."

Two weeks later, I got a phone call from my brother-in-law. I asked, "How is your esophagus?"

"It feels all cured."

"How much medicine do you have left?"

"It's all finished, and I stopped taking it. I only watch my food intake now."

"Two more things: don't drink ice water, and don't drink soda. Wait until you are entirely back to normal."

About a month later, I called. At that time he wasn't at home, but my sister answered. According to her, my brother-in-law was all recovered from that problem and was a happy man.

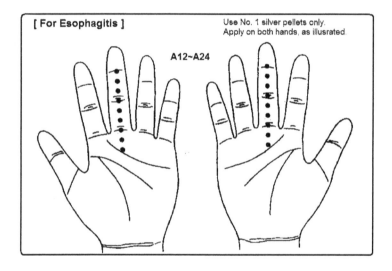

54

Curing a Minister with Chronic Sciatica

THE REV. AHN WAS THE main speaker for a special session we had at our church. He was a very famous minister in our denomination. When the session ended, I sat across the table from the Rev. Ahn while we enjoyed refreshments. All of a sudden, he asked, "Elder Oh, I heard a rumor that you cure lower back pain. Is that true?"

"Yes, but you heard that rumor all the way in Toronto? True, it's relatively easy for me to cure a lower backache and other difficult illnesses."

"Elder Oh, would you fix my sciatic pain, please?"

"How long have you been suffering with that?"

"My sciatica is at least twenty-five years old."

"It hurts a lot?"

"I can't find the words. It's that bad. I'm very sore. I'm numb and losing strength. It's absolutely unbelievable. When

I was preaching my sermon half an hour ago, the pain in my right leg was so unbearable that I couldn't stand on it. Fortunately, my gown goes to the floor, so I could lift my right leg and stand on my left, but I couldn't take a painkiller in the middle of my sermon; I had to sweat it out."

"Reverend, what's your schedule tomorrow?"

"I'm flying to Calgary in the afternoon. I'm leading an evening meeting in the church there and then will fly home straight from Calgary."

"Can you come to my house at nine o'clock tomorrow morning?"

Right at nine o'clock, the Rev. Ahn arrived at my house. I led him to my clinic and asked him to remove his dress pants and lie facedown on the examining table.

The treatment was divided into two steps.

1. I examined the lumbar vertebrae from L1 through to the tailbone. With overlapped thumbs, I pressed on the vertebrae one by one. The Rev. Ahn felt pain at L3, L4, L5, and another few acupuncture points in the area. I marked with a pen everywhere that he responded to the pain. With Genesen, I treated each vertebra for two minutes. I treated the other marked spots for one minute each. Then, I put No. 10 magnets on each marked point.

2. I searched for pain response on the "I" meridians on the back of both hands. I searched from I15 through to I24 and marked wherever I got a pain response.

I marked three points on each "I" meridian and also at M17, M18, and M20. With the black and red leads of the electronic beam on matching points across the middle metacarpal, I treated each pair for thirty seconds. I treated M17, M18, and M20 for twenty seconds. I then placed No. 1 silver pellets on each marked spot on his hands.

I asked him to stand facing me from about two meters (6 ft.) away with his feet about thirty centimeters (1 ft.) apart.

"Raise both of your hands. Now, gradually bend from the waist until your hands are on your knees. Do you have any pain?"

"No."

"Straighten up again and raise both hands to shoulder height at your sides. Slowly turn to the left about ninety degrees. How is your lower back? Any pain?"

"No."

"Twist to the right. How is your back?"

"No pain. Elder, my lower back seems to be relaxing!"

I asked him to stand by the table and, holding the table with his left hand, to raise his right leg.

"Do you have any pain?"

"No."

"Bend your knee and straighten it a couple times. Any pain?"

"No."

"Stretch your right leg behind you as far as you can stretch it. Does that hurt?"

"No."

He repeated the same process with his left leg and discovered that he had no pain anymore.

"Elder, I'm very surprised. My legs were so heavy, but now they feel so light. I'm sure I can walk much faster now. It feels very good."

"Reverend, your illness is so old that you won't easily return to a healthy state. You should have one or two more treatments."

"Yes, I understand. I'm very happy that you helped me and look forward to meeting you again."

The Rev. Ahn continued his travels and headed off to Calgary. That night, he called. "Elder, this morning you helped me out. It is so good, but I have one favor to ask."

"Please, tell me what you want."

"Originally I was to fly back from Calgary to Toronto tomorrow, but I want to change my schedule to return from Edmonton so I can fit in one more treatment from you. Can you do me that favor?"

"Yes, I'll do that, but if you change your plane ticket, it will cost you a lot. Is that okay with you?"

"Money is not a problem; I have to live first! I'll be at your house at ten o' clock in the morning." He had great expectations for my treatment.

The Rev. Ahn arrived at ten o'clock, as promised.

"Okay, have you had any sciatic pain since my treatment?"

"No, it's been very good."

I gave him the same treatment as the day before.

"I didn't know you had such a wonderful talent. I got a lot

of grace from you and don't know how to thank you for all your favors."

The Rev. Ahn left my home smiling broadly.

I was so happy that I could relieve the pain of a fine eighty-year-old gentleman.

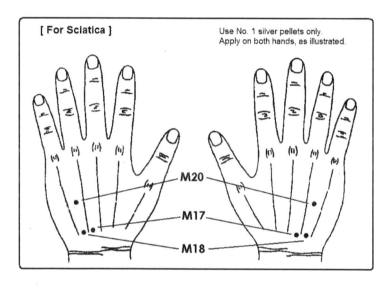

55

Curing a Patient Who Can't Stand Straight Because of Sciatica

A YOUNG WOMAN NAMED KIMBERLY came to visit me.
"Hello. Can I help you?"

"I have sciatic pain that I have suffered from for a long time."

"You don't even look thirty! How can you have such bad sciatic pain? Sciatica is more common among the elderly. How old are you?"

"I'm twenty-seven. My lower back started hurting two years ago, and now my right leg is so sore that I can't stand straight."

"Have you had any treatment in that time?"

"Yes, I went to see many doctors, and I've been getting weekly treatments from a physiotherapist for the past six months."

"Please step over there, turn around, and stand still."

Kimberly's body was leaning to the right, and she couldn't stand straight.

"How did you find me?"

"I flew in from Vancouver for a friend's wedding. I'm just here for a week. I got my hair done yesterday, and one of the hairdressers noticed that I couldn't stand straight. She said, 'Kimberly, go to Dr. Oh right away. He's a magician. A lower backache is nothing to him. He can fix you up right away.' Another hairdresser said she had had a lower backache for three years. She got a treatment from you a year ago and her backache never came back. Then they gave me your address and phone number."

I asked Kimberly to remove her lower clothing and lie facedown on the examining table. I checked her vertebrae, pressing with overlapped thumbs one by one from L1 to her tailbone. She complained of pain at L3, L4, L5, and at a few acupuncture points nearby. At every pain point, I put a pen mark. I treated each lumbar vertebra with Genesen for one minute and the acupuncture points for thirty seconds each. I placed No. 10 magnets (with the N side toward her skin) at every marked point on her back.

I suggested that she stand facing me about two meters (6 ft.) away.

"Put your feet about thirty centimeters (1 ft.) apart and raise both of your hands. Gradually bend from the waist until your hands are on your knees. Do you have any pain?"

"No."

"Now, straighten up again and raise both of your hands

to your sides at shoulder height. Slowly turn to the left about ninety degrees. How is your lower back? Any pain?"

"No."

"Twist to the right. How is your back?"

"No pain. Doctor, I can feel my lower back relaxing!"

I asked her to stand beside the examining table with her left hand on the table for balance.

"Raise and lower your right leg. Now bend your knee and straighten it. Any pain?"

"No."

"Stretch your leg back as far as possible. Does that cause pain?"

"No."

We repeated all of these tests to check her left leg.

"Is that painful?"

"No! This is amazing. I never expected to get rid of this pain so quickly. I've never felt so light and comfortable. My legs move easily."

"Stand straight now."

She stood straight.

"Kimberly, can you come the day after tomorrow for another treatment? If your back has bothered you for so long, there's a possibility that the pain will come back. I recommend at least one more treatment."

Kimberly came for a second treatment two days later. I checked her vertebrae and found she still had some painful points when I pressed. The second treatment was the same as the first.

She made an appointment for a third treatment but called the next day.

"Doctor, I feel like I'm completely healed, so I'd like to cancel my third appointment. I'm leaving for home tomorrow."

"Kimberly, I'm glad you're all cured. Be careful with your body and have a happy life. Goodbye."

I was satisfied to take care of another difficult illness.

56

Curing a Mounted Policewoman Who Had Often Fallen from Her Horse

ONE DAY, A WELL-DRESSED WOMAN in her late thirties came to visit me. Her posture was very straight. Even though she wasn't wearing a uniform, I thought that she was probably police or military personnel.

"You're Dr. Oh? How are you? My name is Marian, and I'm from Yellowknife." (Yellowknife, the capital of the Northwest Territories, is about 1500 kilometers (930 miles) north of Edmonton.)

"Hello. You came from far away!"

"I hear you're good at fixing lower back pain. I've hurt my lower back several times and am suffering with pain."

"You said you've hurt it more than once. How can that happen?"

"I fell from horseback."

"You must be riding a horse often that you have fallen so many times. Is that true?"

"It's an occupational hazard."

"Well, you must be part of the mounted police, right?"

Instead of answering, she nodded slightly.

"Have you had back surgery?"

"No. My doctors have told me that without an operation, my lower back will never heal completely, but I don't want to go through an operation. I'm looking for someone who can fix it without surgery. I've asked many people, and someone mentioned you. Please take care of my lower back."

"I'm glad you didn't go through the operation. If you go through surgery, the doctors usually cut a lot of tissue. Because of that, complete healing takes a long time, and you'd need rehabilitation, too. Can you undress and lie facedown on the examining table?"

I pressed each vertebra down her backbone from T3 through to the tailbone with both of my thumbs overlapped. She complained of pain at T11, T12, L1, L2, L3, L4, L5, and S1. I put a small mark at each spot. In particular, I noticed that the vertebrae at L2, L3, L4, and L5 were tilted to the right.

"Are you having numbness in your right leg?"

"Yes, I am. My right thigh often goes numb."

"This proves that your sciatic nerve is being pinched."

The treatment was divided into two stages.

1. I placed the blue Genesen at T11 and the red Genesen at S1. At that moment, Marian muttered a strong, low "Ugh." I treated her with Genesen for

two minutes. Then, I treated each marked vertebra for one and a half minutes. After the Genesen treatment, I placed No. 10 magnets at each marked point.

2. I looked for the pain-sensitive points on the "I" meridians on each hand. Using the probe from I15 to I24, I checked very carefully and found four painful points. Using the electronic beam, with the black lead and red lead at opposite points across the middle metacarpal bone, I treated each pair for twenty seconds. I placed No. 1 silver pellets on each marked point on the "I" meridians, eight on each hand.

After treatment, I asked her to stand facing me about two meters (6 ft.) away, with her feet about thirty centimeters (1 ft.) apart.

"Raise both hands overhead. Now, gradually bend from the waist until your hands are on your knees. Do you have any pain?"

"No."

"Now straighten up and raise both hands to your sides at shoulder height. Slowly turn to the left about ninety degrees. How is your lower back? Any pain?"

"No."

"Twist to the right. How is your back?"

"No pain."

I told her to leave the pellets on her hands for at least three days and to leave the magnets on her back for four or five days.

She came back one week later for a second treatment. When I checked her back, I saw that all the magnets on her back were still in place.

I was curious. "Why didn't you remove the magnets?"

"I was afraid to remove them for fear the pain might come back after I removed them. By the way, last week, when I flew back to Yellowknife, I had to go through airport security. While I waited in line, I suddenly realized that I had the magnets on and that they might trigger the alarm. I was really concerned. I didn't want to have to undress in public, but there was nothing I could do. I waited and worried in line; there was only one way ahead. When I passed through the metal detector, the alarm didn't go off. What happened? At least I was not embarrassed in front of all those people, but Doctor, is the magnet something special?"

"Yes, the magnet is ceramic, not metal, so the metal detector didn't go off."

"Is that so? I was so worried!"

"Sorry, I never thought of that. By the way, how is your back? Still hurting?"

"No, it seems to be all cured. I didn't ride last week, so I'm not too sure, but the pain was no problem. I feel totally cured. The numbness in my right thigh has disappeared."

"While you're here, what do you think of taking another treatment?"

"Yes, of course."

The second treatment was the same as the first.

After the treatment, I told her, "You can remove the magnets after four or five days. After you take them off, don't throw them away. Those are permanent magnets. Keep them, and the next time you have pain, find the most sensitive painful point and place the magnet there with the N side down. Hold it in place with some adhesive tape. If that still doesn't solve it, come back any time."

Marian thanked me with a smile and left.

This treatment is my patented treatment, so I cannot give it in any more detail.

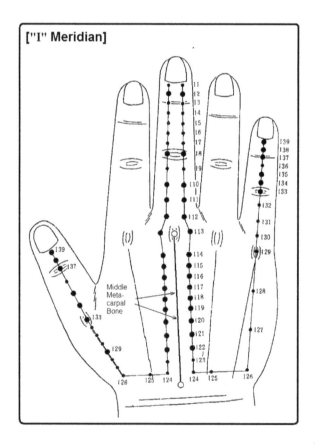

57

A Patient Who Had Had Three Lower Back Surgeries

Mr. Boyko is a high school industrial arts teacher specializing in automotives, but he has a special interest in the rodeo. He's had a number of accidents falling off bucking horses, which hurt his lumbar spine. Consequently, he had three lower back surgeries. According to him, in the last operation, the doctor inserted a piece of metal. He asked me to touch his back, and I could feel the sharp end of the metal. Even after all that, he still couldn't move his back well and couldn't lift any heavy objects.

"Can you fix this kind of lower back problem?"

"I have to try. Once you have had surgery, you lose so many tissues and many important ligaments, and other objects have to be cut off, so it is very difficult to cure. I'm willing to try my best."

"Please do so."

Treatment was divided into two stages.

1. I asked him to remove his shirt and lie facedown on the examining table. I checked the vertebrae from T9 all the way through to the tailbone by pressing down on each vertebra with overlapped thumbs. I put a mark wherever he indicated pain, which happened at T12, L1, L2, L3, L4, L5, and S1. I put the blue Genesen on T12 and the red Genesen on S1. At that very moment, he yelled loudly, "AH!" I gave him two minutes of Genesen treatment and then put No. 10 magnets on each of those seven points.

2. I tried to find the pain-sensitive points on his "I" meridians.* I looked from I15 to I24 on both "I" meridians on each hand. I marked the four spots on each meridian where he indicated pain. Placing the black lead and the red lead of the electronic beam across from each other over the middle metacarpal bone, I treated each pair for twenty seconds and then placed a No. 1 silver pellet on each point.

I asked him to stand facing me about two meters (6 ft.) away, with his feet about thirty centimeters (1 ft.) apart.

"Raise both hands overhead. Now gradually bend from

---

* For "I" meridian, see previous chapter or appendix D.

the waist until your hands are on your knees. Do you have any pain?"

"No."

"Now straighten up and raise both hands to your sides at shoulder height. Slowly turn toward the left about ninety degrees. How is your lower back? Any pain?"

"No."

"Twist to the right. How is your back?"

"No pain. Doctor, this is amazing. I've never felt my lower back so relaxed and moving so easily. By the way, did I scream? I'm sorry—I didn't mean to."

"You don't have to worry about that. I don't think anyone passing the house heard your scream! But your lower back seems to be cured now. Because it's been such a long time, the pain might come back. Why don't you come for another treatment in a week?"

Mr. Boyko came back a week later.

"So, how has your back been this past week?"

"Doctor, I had a big accident. One day when I was teaching, the engine block we were discussing in class broke into pieces.

"The block was hoisted up above eye level. In the middle of my lesson, it dropped, and without thinking, I reached out and caught it. The students were all taken aback, asking, 'Mr. Boyko, are you okay?' They all know I can't lift anything heavy. But I didn't feel any problem. I was really surprised: my back was okay. All the students were astonished that I could catch such a heavy object. This was after only one treatment. I couldn't believe it!"

Mr. Boyko had a second treatment and then left my office with repeated thanks.

This treatment is my patented treatment, so I cannot give it in any more detail.

58

Curing Chronic Lumbago in a Truck Driver

A WELL-BUILT MAN CAME TO my clinic. "Hello, are you Dr. Oh? My name is David. I've been told that you can fix lumbago."

"Welcome. Have you had lower back pain for long?"

"Yes, almost ten years. One day when I was loading heavy objects in my truck, I felt a strange click in my back, and no matter what I do, the pain hasn't gone away."

"What kind of a job do you do?"

"I drive a semi, specializing in long-distance driving."

"What do you mean by long distance? How much mileage are you talking?"

"From the east coast to the west of Canada and back again."

"Wow, that's quite heavy labor. Why don't you take off

your shirt, loosen your belt, and lie facedown on the examining table?"

I started pressing on his back from the midpoint down to the tailbone, looking for a pain reaction. He complained of pain at T12, L1, L2, L3, L4, L5, and S1.

Treatment was in three steps:

1. I used Genesen to treat each painful vertebra.

2. I placed No. 10 magnets on the seven points I found.

3. I looked for pain-corresponding points on his "I" meridians* from I16 through to I24. At every pain response, I put a small mark and then used the electronic beam for thirty seconds on each point, always putting the black and red leads across the middle metacarpal bone from each other.

I asked him to stand facing me about two meters (6 ft.) away, with his feet about thirty centimeters (1 ft.) apart.

"Raise both hands overhead. Now gradually bend from the waist until your hands are on your knees. Do you have any pain?"

"No."

"Now straighten up and raise both hands to your sides at shoulder height. Slowly turn to the left about ninety degrees. How is your lower back? Any pain?"

"No."

---

* For "I" meridian, see chapter 56 or appendix D.

"Twist to the right. How is your back?"

"No pain. I don't have any painful spots anymore. Doctor, my lower back feels unbelievably comfortable and pain free!" He was very happy.

I gave him the following instructions

1. Keep the pellets on the back of your hand for three days.

2. Keep the magnets on your back for four days, and then remove them.

3. Never sit on stone or concrete.

4. Never expose your back to cold winds; always protect your lower back with warm clothing.

Then I told him, "You've had your lower back problem for a long time, so you need more than one treatment. I want you to make an appointment for another visit."

"Doctor, I can come back in two weeks. I'm heading west now and have to drop off a load and then reload and take the new load all the way to Newfoundland. It will take exactly two weeks to return to Edmonton."

Two weeks later, David came at the appointed time. I asked him first, "How has your lower back been so far?"

"Doctor, my lower back is all healed. For the past two weeks, I've had no lower backache. I've never had such a pleasant drive in my whole life. But since I'm here, I'll take your treatment again."

He took off his shirt and lay facedown on the examining table. I noticed that every magnet from two weeks ago was still in the same spot, but with new adhesive over each one.

"Why didn't you remove the magnets?"

"The pain had completely disappeared. I was so comfortable that I worried about what would happen if I removed them. I didn't want to touch them."

I took off all of the magnets, which were covered with grubby adhesive strips, and cleaned and dried the skin with alcohol. I looked for painful points, and only L3 and L4 responded to pressure. This time, I just treated those two points with Genesen and placed No. 10 magnets on each point.

On the "I" meridians on the back of his hands, I couldn't find any pain-corresponding points. David said he was very happy and thanked me, leaving with a big smile.

This treatment is my patented treatment, so I cannot give it in any more detail.

59

Curing an Enlarged Prostate in One Treatment

I HAD BEEN INVITED TO deliver a speech at a Korean university. The day we arrived, the president of the university invited my wife and me for dinner. At the restaurant, we met Mr. Kang, who had also been invited to this dinner. Mr. Kang is the executive director of a large organization and is well-known in the country. We exchanged business cards and introduced ourselves.

Mr. Kang was very interested in an immigrant's life in Canada, so he asked many questions about Canadian life. I explained that after serving twenty-five years as a university professor, I retired and began studying alternative medicine.

"Now I am looking after many patients. Also, I have developed new treatment methods that I am patenting for healing lower backache and whiplash injury. I'm currently developing a new method for treating an enlarged prostate. I have treated many patients who are suffering with prostate problems, and I'm getting very good effects with this treatment."

Just as we were finishing dinner, Mr. Kang asked, "Do you carry your equipment with you?"

"Yes, everything is so small that I carry it wherever I go. I brought it with me this time, too."

"That's a very interesting story. Do you have a cell phone by any chance?"

"Yes, I rented one at the airport." I showed him my cell phone. "I'm still not too familiar with its use, but I can send and receive calls."

"What's your phone number? I might have to see you later."

I gave him my cell number and a brief overview of my schedule while in Korea.

Three days later, Mr. Kang called. "Dr. Oh, I'm having a problem with an enlarged prostate and have been suffering lately. Can you treat me?"

"Yes, I can treat you, but I'm leaving Seoul for another city the day after tomorrow, so I have only one day to see you—tomorrow. I'm leaving the country without coming back to Seoul."

"Fine. How about tomorrow afternoon? Where shall we meet?"

"Treatment is very simple. I can do it anywhere, but you'll need to undress. Do you have a sofa in your office?"

"Yes, I do."

"When does your office close?"

"4:30 PM."

"How about after 4:30 PM, when everyone has left the office? We'll have privacy. It will take about forty-five minutes."

"No problem. That's a good idea. I'll be waiting for you."

The next day, I arrived at his office at about 4:35. When I entered the office, Mr. Kang was the only one there. He locked the door. I asked him to remove his lower clothing, and I used an alcohol sponge to clean his private parts. I used a Genesen treatment there and on three spots on his spine. I placed No. 10 magnets on the spine and No. 1 silver pellets at A1, A2, and A3 on both hands.

With the treatment ended, I returned to our hotel. The next day, I went on to another city to deliver a presentation before coming home.

Three days later, I called Mr. Kang. "How are you? Would you tell me how you feel after my treatment?"

"Dr. Oh, good to hear from you. After you treated me, I was surprised that all the symptoms disappeared. Before, I was getting up four times a night to urinate. Now, I'm getting up only once. Also, I always felt very heavy in my lower abdomen. That also disappeared. I didn't expect such a complete cure after one treatment."

"Good to hear. If you have any problems, don't hesitate to call. I'll give you a call when I come to Korea again."

Three years later, I had occasion to go to Korea. I called Mr. Kang from the hotel in Seoul. "How is your prostate? While I'm here, do you need another treatment?"

"Doctor, amazingly, the prostate has never caused any

problems. It's always very comfortable. I don't think another treatment is necessary."

I was surprised to hear that—the effect of one treatment lasted three years. It meant that his prostate problem was completely cured.

I am applying for a patent for this treatment method. Therefore, I cannot go into further detail.

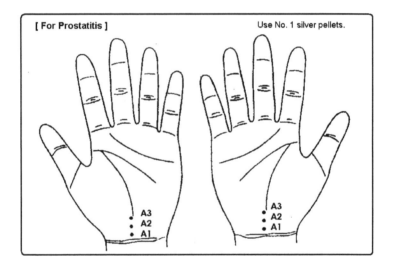

60

A PSA Count that Dropped More than 50 Percent in One Treatment

Mr. Lee is a longtime friend of mine. He is a retired Korean navy officer in his early seventies, but he is very healthy and looks in good shape. On one of my visits to Korea, we had the opportunity to have dinner together. In the middle of the conversation, he said, "By any chance, can you fix an enlarged prostate? I'm suffering a lot these days."

"Of course I can fix that. This is actually one of my specialties. Do you know what your PSA* number is?"

"Yes, seventeen."

"What? Seventeen! It's not cancer? I can't fix cancer."

---

* PSA—prostate specific antigen. This can be tested with a blood test. As the prostate enlarges, the PSA number rises. In the case of cancer, the number will become much higher.

"No, this is not cancer, I'm sure. I got a test about a month ago, and it was negative."

"If it isn't cancer, I can treat you."

The next day, I visited him at his home. Because he had to undress, we used his bedroom, and I treated him on his bed.

I treated his private parts and three spots on the spine. Following the treatment, I placed three pellets at A1, A2, and A3. The whole treatment took fifty minutes.†

After the treatment, we relaxed over coffee at the dining table, and I gave him more instructions.

I asked him to call me the next morning at my hotel because I was leaving the next afternoon.

The next morning, Mr. Lee called and said, "Dr. Oh, that treatment was really effective. I usually get up at least five times in the night. Last night, I only got up once. Earlier, I had worried that the treatment might be painful, but I never felt any pain during your treatment. This morning, my lower abdomen feels very comfortable. This is amazing. When should I take the next treatment?"

"Thank you for the good news. I'm glad you have had such a good outcome from my treatment. It usually gets almost perfect results, and most of my patients never come back for a second treatment. So far, only one has come back for a second treatment, a year later."

"It's unbelievable! When do you come to Korea next?"

"I probably won't have another chance for two or three years. At the moment, I don't have any plans. I have a favor to

---

† I am applying for a patent for this treatment method also and therefore cannot go into more detail.

ask you. Why don't you go to your doctor and ask for a new PSA test? Give me a call and let me know what your new PSA value is. We'll see you next time I visit."

"I'm very thankful. You did such a wonderful job. I promise to send you my new PSA results. Have a good trip home."

As scheduled, we left Korea that afternoon.

Two weeks later, I got a phone call from Mr. Lee. "Hello. I got a new PSA value: it's eight. Doctor, this is unbelievable! In one treatment, my PSA dropped by more than half. It's amazing."

"I never expected it to go down that far. How are you feeling? How many times are you getting up at night?"

"I feel good and get up only once a night."

"Mr. Lee, thank you for such good news. I'll see you next time."

I was amazed. I never expected the treatment to be that effective.

PSA: The diagnostic standard for the prostate specific antigen is

Age	PSA
40s	<2.5
50s	3.5
60s	4.5
70s	6.5

61

A Miracle for a Liver Cancer Patient

Helen is a sixty-six-year-old Korean who had run her own Vancouver business for a long time. One day, she had a strange pain in her upper stomach. She went to a doctor and was diagnosed with cancer of the spleen. The doctor recommended surgical removal of the cancer, and she had the surgery at a famous hospital on the eastern seaboard.

After the doctors removed the affected spleen, within months, they discovered that the cancer had spread to other organs—the uterus, fallopian tubes, and ovaries. Consequently, they removed those organs. Soon after, they discovered cancer in the lymph nodes on both sides of the pelvis. They removed those as well. They found that the cancer continued spreading and had reached the diaphragm. They stripped the diaphragm and the peritoneum.

Helen called me when she spotted an advertisement for

the Korean edition of this book and asked where she could get a copy. With my instructions, she found the book and read it. She wanted to meet me, but said she wasn't well enough to fly to Edmonton.

At the same time, I had occasion to go to Vancouver to look after another patient who was suffering from severe rheumatoid arthritis; the patient couldn't move around her home without someone's help, so she asked me to come and treat her. I took the opportunity to go to Vancouver, and I called Helen and told her I was coming; she could come to my hotel room for an examination and diagnosis.

When I examined Helen, she was quite weak. She explained her problem in detail and reported that she'd now been diagnosed with liver cancer. She brought me a nineteen-page history of her cancer treatments at the eastern hospital and her recent CA-125* (cancer-antigen) test. I briefly skimmed through the reports and noticed that her CA-125 test results had been gradually increasing in the past five months. At that time, her CA-125 value was ninety-two.

I examined her and found that her liver was so overactive in relation to the gallbladder that it was at a near-critical stage. The treatment plan I explained was as follows:

1. Drink special soup (see appendix A).

---

* CA-125 tests the cancer cells in the blood by measuring the antigen value. In other words, the greater the number, the more cancer cells are in the blood. If the number is less than twenty-six, we interpret it as safe.

2. Take one-eighth of a tablet of *pien-tze-huang* twice daily.

3. Place pellets on both hands according to the *liver-excessive* formula.

4. Spinal treatment at T4, T5, T6, T7, and T8 with two minutes of Genesen on each vertebra.

5. Wear a therapeutic ring on each thumb.

6. Watch your diet: no fatty, spicy, salty, or hot foods, and no alcohol. Control your emotions so that you never get excited, angry, or depressed.

7. Pray to God every day for help.

8. Get a CA-125 test done every month.

Helen called me a month later. She reported that her new CA-125 test result was seventy.

"Doctor, dropping twenty-two points in the CA-125 test is almost impossible in the traditional understanding of medicine. My family doctor was amazed and wants to know what kind of treatment I've received. My doctor says she wants to see you when you come to Vancouver next time."

I treated her once a month in visits to Vancouver, and every time, her CA-125 result continued to drop, from seventy to sixty-two, to fifty-five, to forty-four. When the CA-125 had dropped to forty-four, she said she was strong enough to fly, so she came to Edmonton to get her treatments.

I suggested one more medicine in her treatment: the plant graviola grows along the Amazon River and is known as a natural cancer killer. One study shows that graviola is capable of killing cancer cells at a rate ten thousand times stronger than commonly used chemotherapy drugs.

I suggested that Helen use N-Tense tablets, which contain 50 percent graviola essence, but somehow she got the actual graviola fruit itself (sent frozen from a relative in Miami). I suggested that she take N-Tense because its controlled production includes tests for toxicity, but she refused to take the pills and preferred to take graviola itself. I was not too happy to see her taking raw, frozen graviola because I had no idea what the side effects might be.

At the time of this writing, Helen's CA-125 had gone up about ten points after a very difficult family situation during which she was very emotional. I conclude, however, that as long as she's taking my advice, she should soon recover from her cancer.

62

A Stubborn Terminal Liver Cancer Patient

Mr. Shin, sixty-seven, used to run a laundry facility in Toronto. For several years he had suffered with liver cancer and was finally admitted to the hospital. One August 9, his doctor told him that the cancer cells were now affecting 95 percent of his liver; there was no cure, and he probably had only one month to live. His doctor suggested that he go home, gather his family together, and enjoy his last days. "Leave a good impression before you pass away," he advised.

Mr. Shin was discharged and went home, prepared to die. He lay in bed waiting for his final moment.

One of his friends, Mr. Kim, read my book with interest, particularly chapter 2. Mr. Kim called Mr. Shin about the book and recommended that he contact me, but Mr. Shin replied, "My time is up anyway, so that's useless."

Mr. Kim called me directly and explained the situation.

He asked my opinion. "Is it really too late, or is there any way he can survive?"

I answered, "If he starts taking the special soup [appendix A] right away, there's no reason he couldn't survive. Tell him to call me directly so I can ask specifics about his symptoms and give him instructions."

The next morning, Mr. Shin called. "Dr. Oh, I'm Shin. You heard my situation from Mr. Kim. Do you think I still have a chance to live?"

"Yes, Mr. Shin. You still have a chance. Don't give up. Read my Korean book carefully, particularly pages 26 and 27, and follow the recipe to make the special soup. Ask your wife to do it immediately, and starting this evening, drink two cups a day: one in the morning and one in the evening. Can you get *pien-tze-huang?*"

"I know that medicine. I still have about three tablets."

"Good, break each tablet up into eighths and take one portion once in the morning and once in the evening with warm water. Trust me; believe in God, and pray. You'll be fine. Don't worry."

"Yes, I'll do that. I'll ask my wife to prepare the special soup right away."

Two days later, I got a phone call from Mr. Shin. "Dr. Oh, I started taking that soup the same day I called you, and I'm feeling much stronger and feeling more comfortable. I'm getting well?"

"Yes, you're getting a good effect. Just keep it up until I come and treat you directly."

"Doctor, when are you coming?"

"My flight is scheduled for October 5."

"Doctor, that's too late. I'll be gone by then."

"Mr. Shin, if you take the special soup and medicine as I directed, you won't die. Don't worry. I'll meet you in Toronto. My flight arrives at two, and I'll be checking in at the hotel by four o'clock. If you can make it, I'll meet you there."

"Dr. Oh, I'm very weak, so you'll have to come to my house to treat me."

"If you're weak, of course I'll come to your place. Don't worry. You'll get the treatment."

When I arrived at my hotel, Mr. Shin and his wife were waiting for me in the lobby. I immediately noticed Mr. Shin because his face was quite dark (typical of liver patients). As soon as we greeted each other, both of them were asking, "Can Mr. Shin survive?"

I assured them, "No problem. Let's go up to my room."

In the room, I examined him just as I had Helen (chapter 61) and applied pellets on both of his hands according to the *liver-excessive* formula. I then treated his vertebrae with Genesen and gave him two therapeutic rings, one for each thumb. I gave him the same instructions that I had given Helen and asked him to come back two days later for a second treatment.

The next day, I got a phone call from Mr. Shin.

"Dr. Oh, I'm very healthy and strong and don't feel any further treatment is needed. I'd like to cancel tomorrow's appointment."

"Are you sure you don't want a second treatment? This is very important."

"No problem. I'm very healthy now. I don't need it. By the way, I understand you're going home on Friday evening. I'd like to give you a ride to the airport."

"You, a terminal liver cancer patient, want to drive me to the airport? Are you sure? Can you make it?"

"Sure. I'm fine, Doctor. No problem at all."

"Well, if you insist, I'll take a chance. Please pick me up Friday afternoon and give me a call for what time."

I figured that if he could not drive all the way to the airport, then I could take over, so I wasn't worried about that. Still, I was not too happy about his stubborn refusal to take a second treatment.

Mr. and Mrs. Shin came to pick me up. It took two hours through the rain on one of the country's busiest freeways to reach the airport. I was amazed at how well he had recovered. He never showed any fatigue all the way to the airport. Even in the car, he was telling me that because he was fully recovered, he wanted to start a new business. I was flabbergasted at that comment.

"Mr. Shin, next week please go to your family doctor and ask for a CA-125 test and let me know what the new value is. If you have previous records of your CA-125, give me those values, too."

I returned home safely that night.

A few weeks later, Mr. Shin called. He reported that his previous CA-125 was 138, but the new value was 33. His doctor was amazed at such a decrease from one treatment. I was also surprised. At the end of our conversation, he said, "Doctor, I'm all cured, right?"

"Mr. Shin, I don't think you're completely cured yet, so follow my instructions carefully, and never omit anything."

About two months later, I got a phone call from Mr. Kim. "Dr. Oh, Mr. Shin called me yesterday. His abdomen is swelling and both legs are swelling, too. According to his wife, he took a trip somewhere for two weeks and returned home in that condition."

I asked Mr. Kim to have Mr. Shin call me directly so I could hear his exact symptoms, but Mr. Shin refused to call me. I asked Mr. Kim if he could get me the value of Mr. Shin's latest CA-125 test.

About a week later, Mr. Kim called again. "His new CA-125 test value is four hundred." Three days later, Mr. Kim informed me that Mr. Shin had passed away.

I was extremely disturbed by such a result. I have no idea why he refused to take the second treatment or how closely he followed my instructions.

63

Instantly Raising a Friend Who Had Been Bedridden for Forty Days

Mr. Lee, a well-known novelist, is a close friend of mine. After I returned from one of my trips to study alternative medicine in Korea, I called him to set up a lunch meeting.

"Mr. Lee, I just came back from Korea. Why don't we get together for lunch today?"

"Dr. Oh, I'm sorry. That's a good offer, but I cannot go."

"What's going on? Why can't you come out to lunch?"

"I've been bedridden for forty days. I can't even sit up. I take my meals lying sideways on the bed. I have to crawl to the toilet. It's a hard life."

"No kidding? You must have a lot of trouble. What kind of medicine are you taking?"

"Tylenol 4. I'm taking twelve pills a day."

"Mr. Lee, I'll be at your place in an hour. Just wait."

I arrived with my treatment bag. He looked quite miserable. His wife told me that it really was the fortieth day. They asked if I had brought any good medicine.

I looked for the pain-sensitive points along his "I" meridians on both hands with the probe and found two points on each meridian. I treated each pair with the electronic beam for one minute with the leads on the matching spots on either side of the middle metacarpal bone. Then, I asked him to turn over. First, he said that that would be impossible, but then he realized that his back seemed more relaxed, so he did manage to turn.

With him lying facedown, I found he had pain at T3, T4, and T5 when I pressed. I used the blue Genesen on T3 and the red on T5 and treated him for two minutes. In addition, I treated each vertebra individually for one minute. After that, I placed No. 10 magnets on each vertebra and gave him two therapeutic rings, one for each middle finger.

I told him, "Okay, Mr. Lee. Now you're okay. You can get up now."

"Get up?" He stared at me in disbelief.

I pulled his hand, repeating, "Get up." He started to object, but I pulled him, and he gradually got to his feet.

"Oh, I can stand!"

"Now, let's go to the living room and sit on the sofa."

"Ah, it's wonderful to sit on the sofa for the first time in forty days. I can't believe you fixed it so quickly and easily. Dr. Oh, you did a wonderful thing that I really appreciate. Do you know how difficult it is for a man to lie in bed for forty days?"

His wife was very appreciative, and I was happy to see that I could do something so good for my friend.

64

Prostate Patient Returns for Second Treatment after Five Years

Marco is an Italian immigrant in his midforties. He called for an appointment and came to my office. On arrival, he explained that he needed another treatment for his enlarged prostate.

I asked him, "When did you take my first treatment?"

"Five years ago."

"Did you have any problems during those five years?"

"Not at all. Only recently, I started to get a heavy feeling in my lower abdomen, and I was getting up four times a night."

"Marco, how was it after you got that first treatment?"

"At that time, I was getting up five times a night, but after your treatment, I only needed to get up once a night. I had no problems, and my lower abdomen felt comfortable and light.

Things just changed recently, so I've come for a second treatment."

I treated him as I had five years before and placed No. 1 silver pellets on A1, A2, and A3 on both hands.

Marco went away a happy man, commenting on how surprised he was that one treatment had lasted such a long time.

I am applying for a patent for this treatment method. I regret that I cannot disclose more detail than this.

65

Reversing Hearing Loss

One day, I felt a flood in my left inner ear. Within ten seconds, I lost hearing in both ears. I went to our family doctor for help, and he sent me to an audiologist to get a hearing aid. I thought that was strange because he didn't try to discover why I had suddenly lost my hearing. He didn't even examine my ears.

I went to an audiologist, who tested my hearing and said I needed hearing aids. I discovered that even hearing aids didn't give me back my hearing.

Several months passed. At a religious meeting, a minister gave me a special prayer. The prayer stopped the ringing in my ears, but my general hearing didn't return. It was difficult to understand people's speech, and I always felt like my ears were under pressure or filled with fluid.

I asked my family doctor to refer me to an ENT (ear, nose,

and throat) specialist. When I went to the ENT specialist, he asked, "Were you in a war?"

"Yes, I was."

"That cost you. Did you shoot guns?"

"Yes."

"That's why you lost your hearing."

He didn't even think of other possibilities, so I felt totally helpless.

Two years later, by chance, I was introduced to another ENT specialist who measured my ear pressure. He said, "Your pressure is 0.3. That's almost the same as if you were in an airplane."

He explained that my sinus cavity was swollen and blocking the Eustachian tube between the ears and throat. He prescribed a nose spray to bring down the swelling. I sprayed the medicine as instructed, but the effect was not that great. I found it more effective to take down the swelling using pellets. I placed one No. 6 silver pellet on either side of A28 on the middle finger of both hands. I could recognize someone's voice, but I couldn't understand the words spoken.

Then I saw an article about ear candling. I searched for an ear candling service through the Internet and found one in my city. An East Indian woman administered the ear candling for me the first time.

She asked me to lie on my side with my left ear up. She placed the ear candle in my left ear and lit it. While I was being ear candled, she put massage oil on the side of my head and constantly massaged my cheek, temple, and all around my ear. After the candle had burnt down to about one third of its original length, she put it out.

After she extinguished the candle, she split it open and showed me what had accumulated. My wife, who was observing, and I were really surprised at the amount there—almost a tablespoonful in two different colors: part the color of the candlestick and part the color of earwax. The amount of earwax was about a teaspoonful.

Then she asked me to roll over so she could do my right ear. When she'd put out that candle, she opened it up, and the amount of accumulated wax was about the same.

I asked her, "Is this wax or earwax?"

"The different colors tell: the dark wax is what came out of your ear."

We were surprised at the amount of earwax. The size of the outer ear is very small, but the amount of earwax was so large. There must have been some coming from the inner ear as well.

After this treatment, I realized that I could hear speech much more clearly than before. I've repeated the ear candling about six times now. Every time, the amount of wax extracted is about the same. And every time, the ear candling gives me better hearing and better understanding of words in particular.

Recently, I noticed that my hearing directionality has also improved; that is, I can better distinguish from which direction sounds are coming.

Although there are people who say ear candling is dangerous or that it doesn't work, I am convinced that this method is helping me regain my hearing.

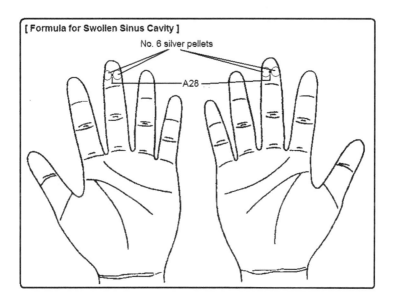

66

Korean Hand Therapy Explained

Trained as an acupuncturist, Dr. Tae-Woo Yoo invented this alternative medicine in the early nineteen-seventies, mostly drawing from his personal experiences and basing it on ancient Oriental philosophy. The illustrations have been copied from Dr. Tae-Woo Yoo's book with the author's permission.

1. Theory of corresponding spots on both hands

    Certain spots on the hands reflect the function of the whole body and the internal organs. As illustrated, 345 spots in each hand (see below and appendix D) correspond to specific organs or body functions. For example, if you have a stomach problem, see that A12 is the stomach corresponding spot. You can acquaint yourself with the corresponding spots for the internal organs and other body parts in the following illustrations (or see appendix D).

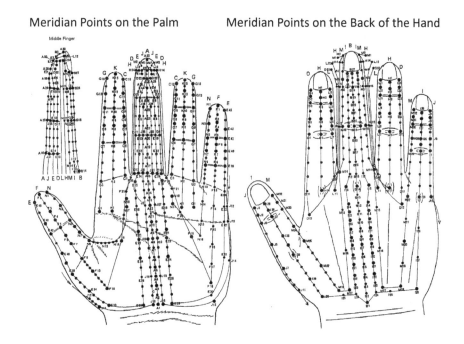

Meridian Points on the Palm   Meridian Points on the Back of the Hand

* Detailed illustrations are available in the Appendix.

Points on the Palm Control Internal Organ

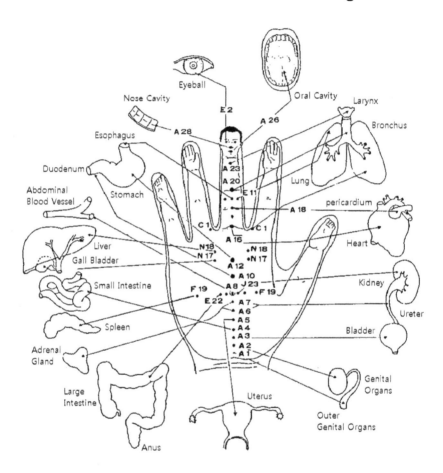

Face and Corresponding Points

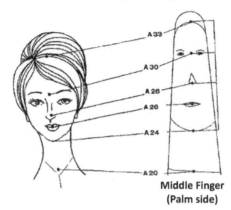

Middle Finger (Palm side)

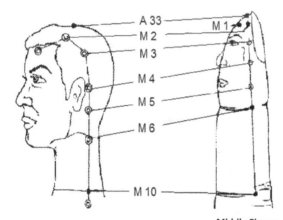

Middle Finger

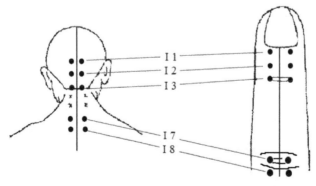

Middle Finger

Corresponding Points Control Pain in the Body

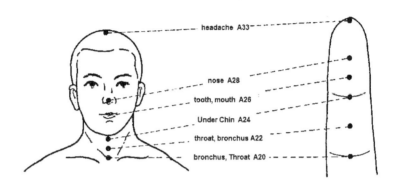

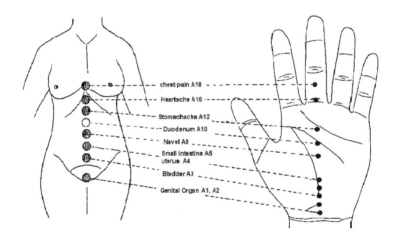

Various Joints and Corresponding Points in the Hands

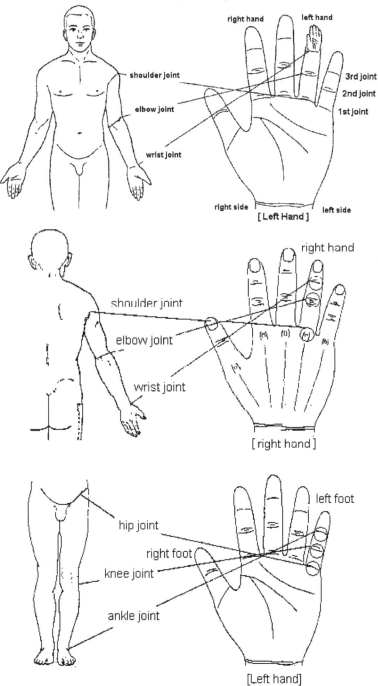

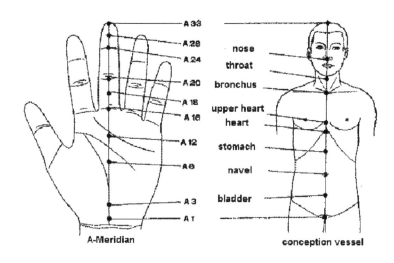

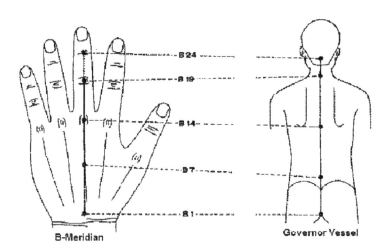

Fingers Correspond with Body, Hands, and Legs

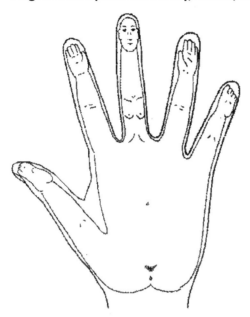

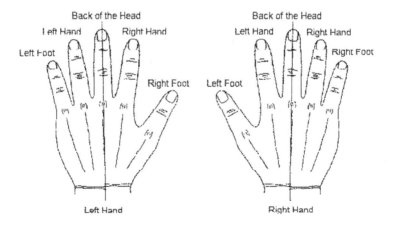

2. Theory of balance between two internal organs

According to his theory, Yoo reclassified the ten organs in our body into five pairs:

- liver–gallbladder
- heart–small intestine
- spleen (and pancreas)–stomach
- lung–large intestine
- kidney (and reproductive organs)–bladder

These pairs are in a yin and yang relationship and should be equal in function and strength. Once the balance is broken, numerous symptoms occur. For example, if the strength of your liver is excessive compared to that of your gallbladder, you might become very nervous and experience insomnia, headaches, vision problems, indigestion, or fatigue, among other possibilities. On the other hand, if your gallbladder were overactive, you might get symptoms like arthritis, migraines, sciatic pain, lower backache, and gallstones. These are only some of the many symptoms one might experience.

To maintain an equal balance, we have to control the energy flow to and from certain organs by using pellets, electronic beams, Genesen Acutouch, or magnetic force.

3. Theory of fourteen meridians on the hands

    As you see in the illustrations (appendix D), there are fourteen meridians in each hand that connect the 345 spots. Each meridian is a route through which energy is supplied to the designated organs. The strength of the energy is controlled by the use of pellets, electronic beams, Genesen Acutouch, or magnets.

4. Theory of relationship between five fingers and internal organs

    Each finger has a controlling function for certain internal organs:

    - thumb controls liver and gallbladder
    - index finger controls heart and small intestine
    - third finger controls spleen (and pancreas) and stomach
    - fourth finger controls lungs and large intestine
    - pinkie finger controls kidneys (and reproductive organs) and bladder

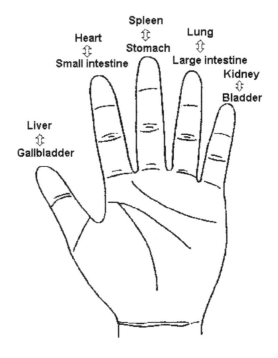

5. Theory of O-Chi formulas

When an imbalanced organ needs to regain balance, we use silver and gold pellets on specific meridians. The organs in the body cannot exist in isolation. We have to think about balance with the other organs, as well.

When you adjust the liver–gallbladder balance, those organs have a close influence on the heart meridian and the lung meridian. Therefore, instead of adjusting the energy flow only in the liver meridian, we need to consider the neighboring heart and lung meridian. The O-Chi formula considers the neighboring organ meridians as well. Therefore, using the O-Chi formula is more effective than adjusting only the meridian of concern.

The O-Chi formula always uses two No. 1 silver pellets and two No. 1 gold pellets. The thirteen most commonly used O-Chi formulas are included in appendix C.

67

Korean Hand Therapy Applied

All the stories revealed in this book are from my own experiences. As I write this book, I can recall my patients' faces, their circumstances, and my own feelings of accomplishment. Sometimes I wondered how I could achieve such cures, but I assure you that all of the stories are true, with nothing embellished. Some may say this is unbelievable, exaggerated, or self-praising. The happiness of the people who are so glad to be cured gives me the deepest feelings of satisfaction.

My way of approaching a patient

1. When a patient makes an appointment over the telephone, I try to get as much information as possible about the symptoms.

2. I read up on that problem. Sometimes this keeps me up late into the night, but I take notes of all the possible treatments. Sometimes one problem can have many different treatments. I study each treatment method carefully and select one according to what would best suit the patient's situation.

3. Every morning in my prayers, I pray for my patients and pray for good judgment in the selection of the proper method.

4. When a patient comes to the office, I ask once again about the specific details of the symptoms. If the patient knows any results of diagnostic procedures from a family doctor or hospital, I ask what diagnosis has been given.

5. Through checking the abdomen and pulse reading, I work to discover the patient's specific health condition, traits, and the cause of the illness to determine my own diagnosis.

6. I select a treatment method and apply the treatment.

7. For neck, back, sciatica, and arthritis patients, I always test the effect of the treatment to make sure it worked and that all of the painful points are taken care of.

8. At the end, I always give instructions about after-treatment care.

Applying the treatment methods presented in this book

You can carry out many of the treatments in this book yourself. If you want to apply the treatment methods, follow these instructions carefully.

1. The treatment method may seem simple but should not be administered carelessly. In particular, when you try to find the pain-corresponding point, it is important to be meticulous.

   When you use a probe, place the probe in one spot, push down, and keep it there for at least five seconds. If that is not a pain-corresponding point, using it as a base, move one millimeter (1/32 in.) along the meridian and test again. Continue to move in very small steps. You won't be likely to find the points by randomly jumping around. Once you find a pain-corresponding point, always mark it accurately with a pen.

2. When you use pellets, remember that silver pellets have a force of pushing energy while gold pellets have a power to collect energy.

   Example: If you have an overactive heart, energy

flow in your heart meridian is moving too fast. Therefore, placing silver pellets at the larger-numbered point and the gold pellets at the smaller-numbered point will produce a slowing or lessening effect. Consequently, the balance between heart and small intestine will be normalized.

If you suffer frequent urination, it means that the bladder is deactivated; in other words, the energy flow in the bladder meridian is too weak. Therefore, when you place silver pellets at the smaller-numbered point and gold pellets at the larger-numbered point, you help strengthen or speed up the energy to balance the bladder with the kidney.

3. When you apply the pellets, you have to be exact so that the center of the pellet rests precisely on that spot.

4. Once you place the pellets, keep them there for at least three days. If you have to keep the pellet in the same spot for a long time, remove it after three days and rest for one day before continuing for another three days.

5. When you use magnets, always place the N side down against the skin.

6. When you use the electronic beam, remember that

electrons flow from the black lead (–) to the red lead (+).

In order to control the flow of energy in the meridian, study a chart of the meridians, and make sure that you know which way the energy flows. An easy way to check is that energy flows naturally from smaller to larger numbers. Therefore, if you put the black lead on the smaller number and the red lead on the larger-numbered point, you are strengthening the energy flow; if you do the reverse, then you are reducing the energy.

The standard treatment time is twenty seconds.

7. When you use Genesen Acutouch, Genesen has to be placed perpendicular to the skin on the spot. Blue Genesen is N, and red Genesen is S. When you affect the flow of energy in the meridian, remember that magnetic force flows from N to S.

My wish is that this small book will help liberate people who suffer from illnesses and give them a happy life.

Appendix A

Life Extension Formula: The Special Vegetable Soup

Dr. Kazu Tateishi's remedy for cancer and other illnesses

When cancer strikes a family, what course of action should one take? Dr. Tateishi's father and brother both died of cancer, and he found himself suffering from cancer of the duodenum and surrounding area. His stomach was removed, but the cancer cells had spread to his lungs. He resolved to fight for his life. He researched, studied, and tested over fifteen hundred types of herbs and plants. Eventually, he discovered the right combination of ingredients and formulated a unique healing vegetable soup and brown rice tea with its own molecular powers. The vegetables are rich in green chlorophyll, amino acids, iron, phosphorous, and calcium, all in a natural form.

He based his soup on the principles of the Five Elements theory, the harmonizing balance of the forces of yin and yang, acid and alkaline, which engenders health as opposed to the imbalance, which leads to disease.

The five elements in our environment are wood, fire, earth, metal, and water. Because of the balance of the five elements, the earth is able to produce life. Dr. Tateishi used the essence of the five elements. Each essence possesses its own color: green, red, yellow, white, and black. These relate to the corresponding internal organs: heart, liver, spleen, lung, and kidneys. He used the concept of the five different colors as matched to specific vegetables:

> Green: daikon radish leaves
> White: daikon radish
> Red: carrot
> Yellow: burdock root
> Black: shiitake mushrooms

What this remedy may do:

- Heal cancer
- Restore injured joints and bone structure
- Slow down the aging process
- Rejuvenate the skin
- Heal cataracts
- Heal liver disorders
- Lower high blood pressure
- Improve heart conditions
- Reduce brain tumors and other related head-injury problems
- Reduce high white blood cell counts
- Improve T-cell function

This is a remedy for all age groups. When the soup is ingested, it produces thirty different elements for fighting diseases. Within three days, it can stop the growth of cancer cells. Even for some last-stage cancer patients, it can lead to 100 percent remission. Patients who have subsisted on feeding tubes for nourishment have regained their strength after two days of being given the soup. Over ten thousand last-stage cancer patients have used this remedy, and 99 percent of them survived and returned to work.

Special Soup Recipe

Ingredients

200 g (7 oz.) white daikon radish (cut into double bite-size; do not peel)
50 g (2 oz.) white daikon leaves (cut into one-half or one-third length)
50 g (2 oz.) carrot (cut into double bite-size)
50 g (2 oz.) burdock root (cut into double bite-size)
1 sun-dried shiitake mushroom
2 liters (8 cups) water

Directions

Don't peel anything, don't add any other ingredients, and don't add any seasoning.

If you are not sure whether the mushroom is sun-dried, expose it to the sun for about one hour.

Use only a stainless steel or glass pot. Never use a porcelain pot or a Teflon-coated pot (coating may melt). With a cover on the pot, use high heat. When the mixture comes to a boil, reduce heat and simmer for one hour.

Strain and drink.

Store the soup in a glass jar and keep it in the refrigerator.

Dosage

Drink 200 mL (7 oz.) at least twice a day to a daily maximum of 600 mL (20 oz.).

Never drink soup that is more than three days old.

You can drink it either hot or cold.

Brown Rice Tea Recipe

Another remedy that can be used in conjunction with the vegetable soup is brown rice tea.

It is a good diuretic for ridding excess water from the body. For diabetics, it can bring the blood sugar level down by producing more natural insulin in the body. It cleanses the blood and blood vessels. Someone with a heart condition drinking both soups for twenty days will be able to regulate their problem by drinking three cups per day. For a regular cancer patient, about two cups per day is sufficient.

Ingredients

1 cup brown rice
16 cups water (divided)

Directions

Roast the rice without oil until dark brown, but don't burn. Boil 8 cups of water and pour roasted rice into water. Turn off heat. Cover and let sit for five minutes. Strain out rice and reserve liquid.

Boil 8 more cups of water. Put the strained rice in this water. Cover, lower heat, and simmer for five minutes. Strain rice, reserving water. Combine the two rice waters.

Special Notes

The amount one needs depends on one's condition. The more serious the condition, the more tea should be drunk.

Do not drink tea while taking a high protein substance.

Don't drink the vegetable soup and brown rice tea together. Wait at least fifteen minutes.

Appendix B

Equipment and Materials

1. Pellets, Moxa, Magnet, Ring
2. Probe
3. Electronic Beam
4. Genesen Acutouch

1. Equipment and Materials

(1) Pellet, Moxa, Magnet, Ring

No. 1 Silver Pellet

No. 1 Gold Pellet

No. 6 Silver Pellet

Moxa

No 10 Magnet

Therapeutic Ring

(2) Probe

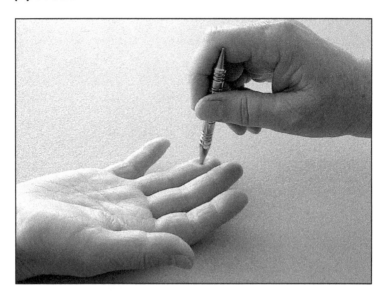

(3) Electronic Beam

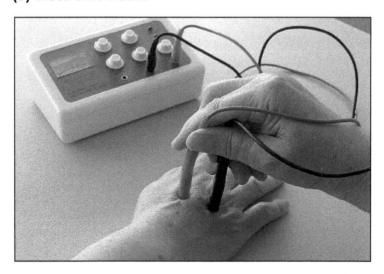

(4) Genesen Acutouch

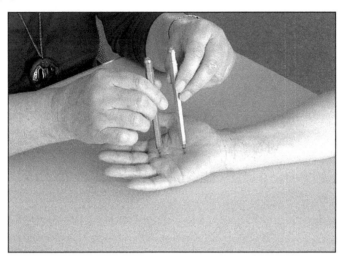

All the materials and equipment demonstrated in this book are available for purchase from

    Korean Hand Therapy Research   1-780-487-3727
    15503 – 76 Ave. NW   1-780-452-3603
    Edmonton, AB T5R 3A4   Fax: 1-780-487-3727
    CANADA   choong_oh@hotmail.com

    Website: www.KoreanAlternativeMedicine.com

Appendix C

Frequently Used O-Chi Formulas

1. Heart-Deficient
2. Heart-Excessive
3. Liver-Excessive
4. Spleen-Excessive
5. Stomach-Deficient
6. Stomach-Excessive
7. Lung-Excessive
8. Lung-Wet
9. Large Intestine-Excessive
10. Kidney-Deficient
11. Kidney-Excessive
12. Basic Formula for men
13. Basic Formula for women

Note: (G) means to place a No. 1 gold pellet at the spot.
(S) means to place No. 1 silver pellet at the spot.

(1) Heart - Deficient Formula

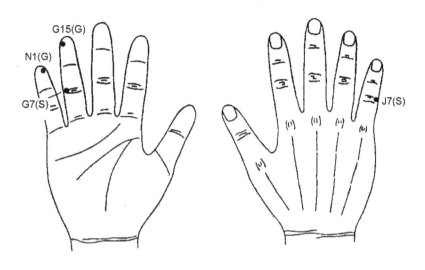

(2) Heart - Excessive Formula

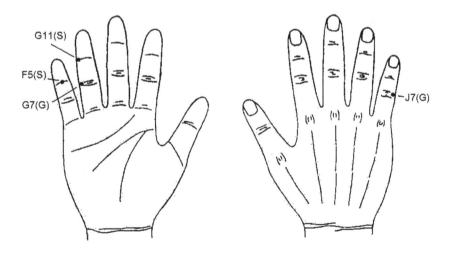

(3) Liver - Excessive Formula

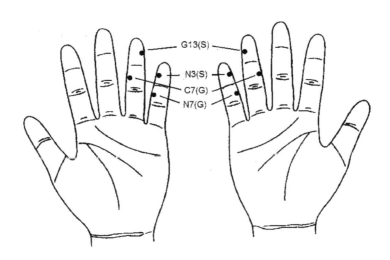

(4) Spleen - Excessive Formula

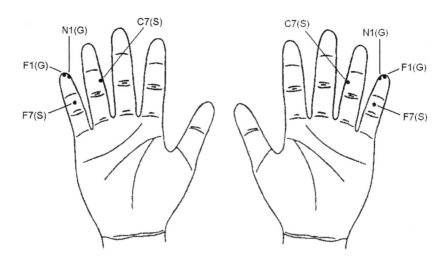

(5) Stomach - Deficient Formula

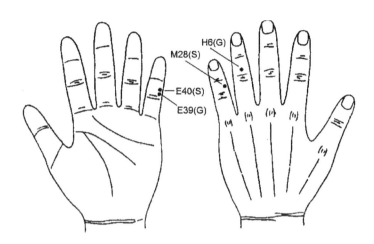

(6) Stomach - Excessive Formula

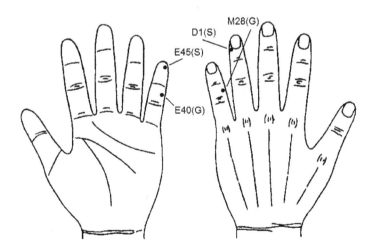

(7) Lung - Excessive Formula

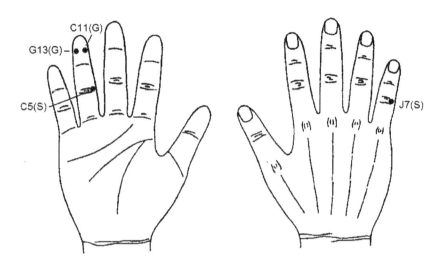

(8) Lung - Wet Formula

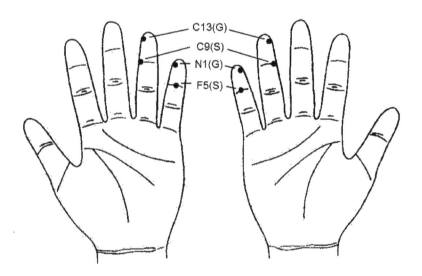

(9) Large Intestine - Excessive Formula

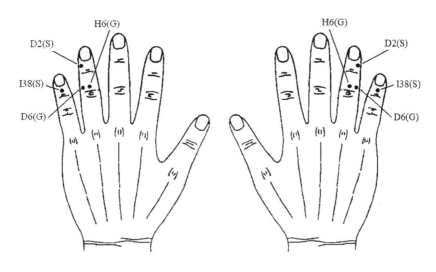

(10) Kidney - Deficient Formula

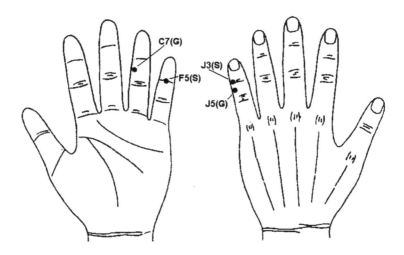

(11) Kidney - Excessive Formula

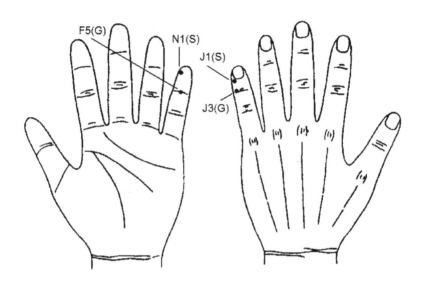

(12) Basic Formula

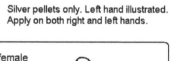

Silver pellets only. Left hand illustrated.
Apply on both right and left hands.

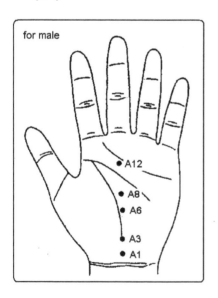

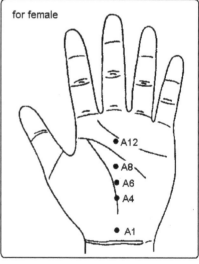

Appendix D

Meridian Charts

The following sixteen charts have been copied from Dr. Tae-Woo Yoo's book with the author's permission.

A-Meridian	G-Meridian	M-Meridian
B-Meridian	H-Meridian	Meridian points on the palm
C-Meridian	I-Meridian	Meridian points on the back of the hand
D-Meridian	J-Meridian	
E-Meridian	K--Meridian	
F-Meridian	L-Meridian	

(1) A-Meridian (Conception Vessel)

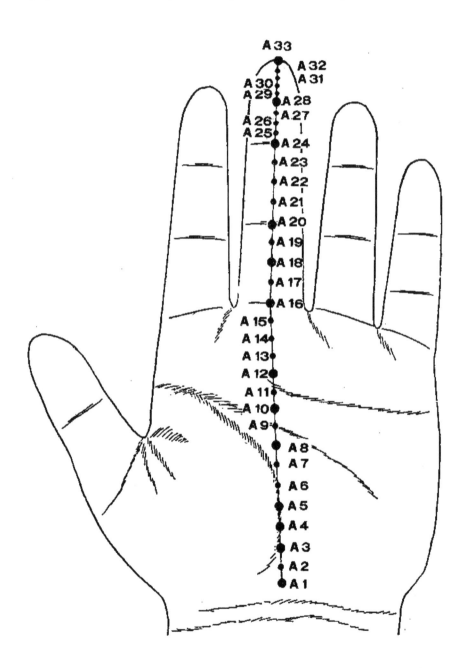

(2) B-Meridian (Governing Vessel)

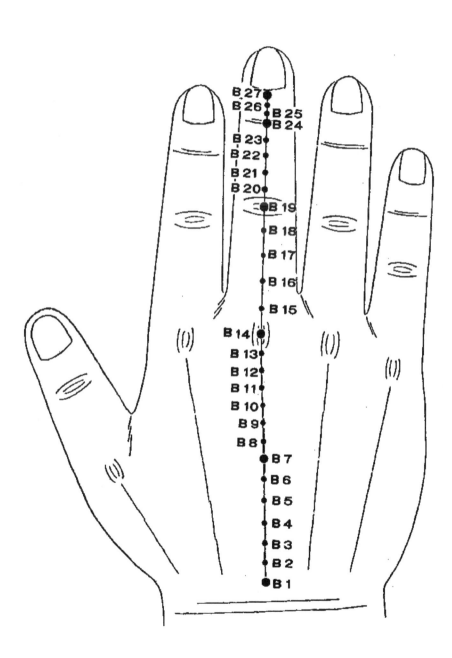

(3) C-Meridian (Lung)

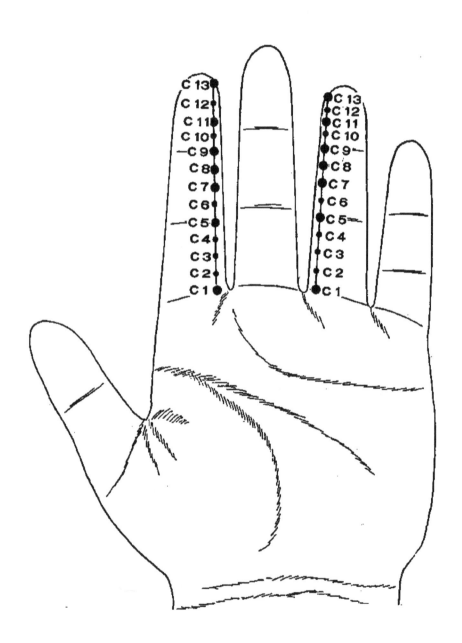

(4) D-Meridian (Large Intestine)

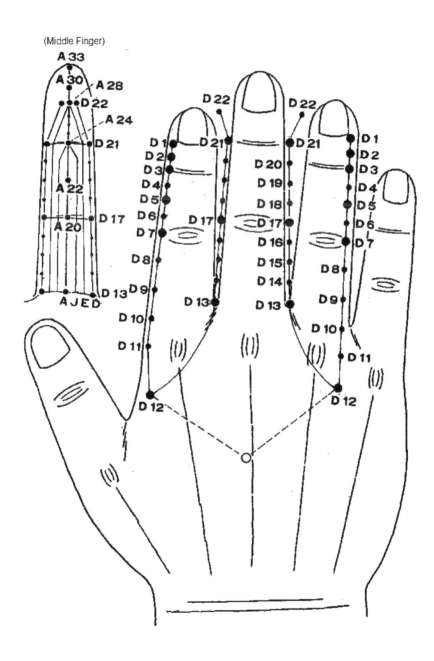

(5) E-Meridian (Stomach)

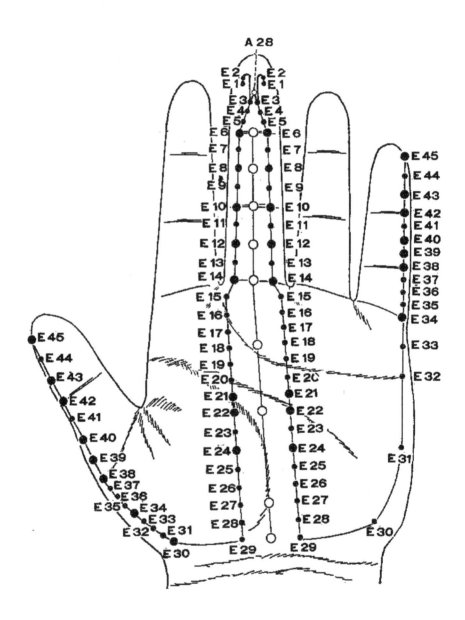

(6) F-Meridian (Spleen)

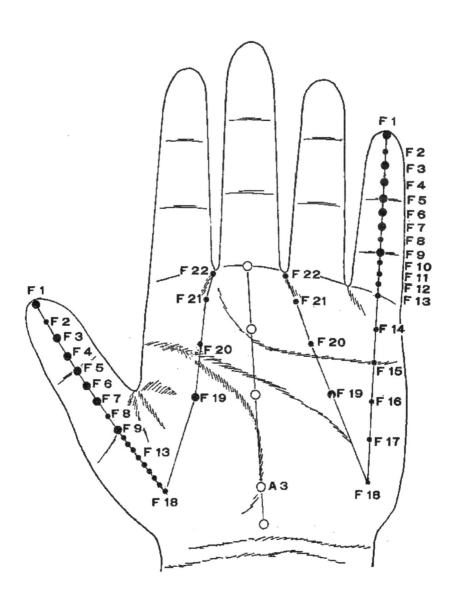

(7) G-Meridian (Heart)

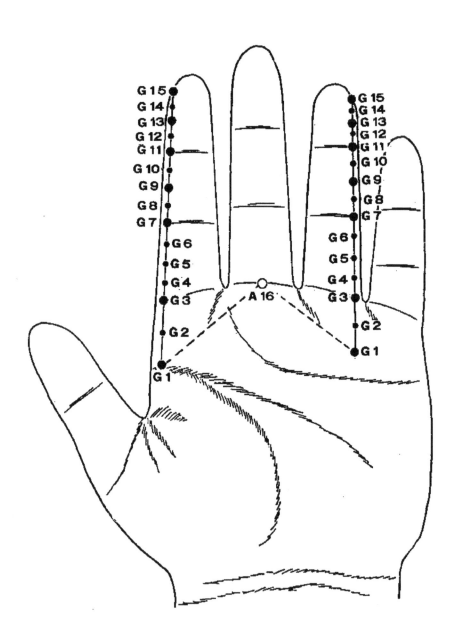

(8) H-Meridian (Small Intestine)

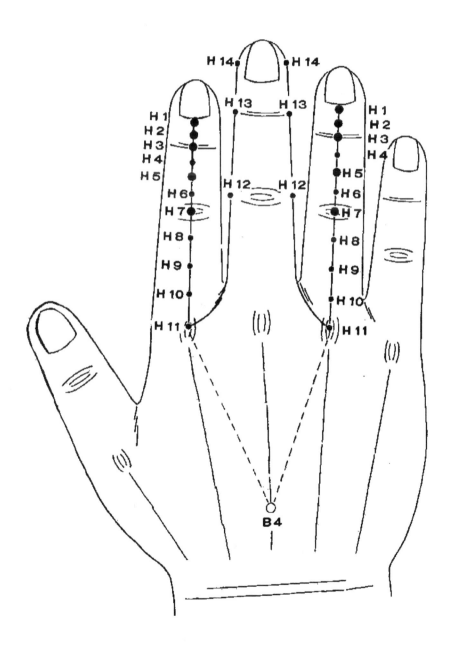

(9) I-Meridian (Bladder)

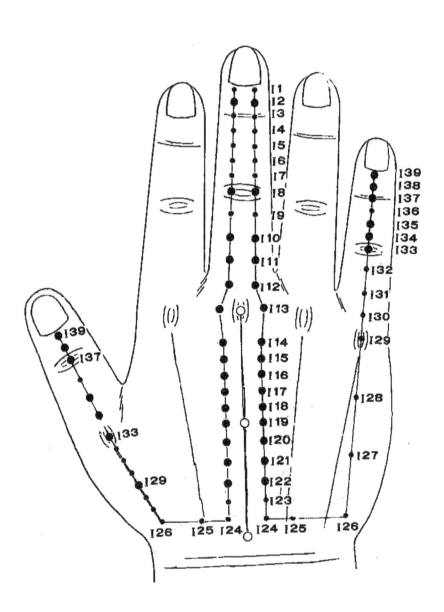

(10) J-Meridian (Kidney)

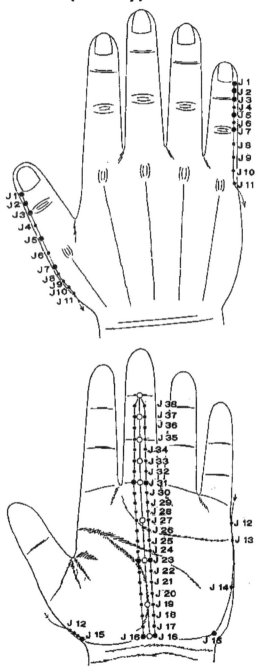

(11) K-Meridian (Pericardium)

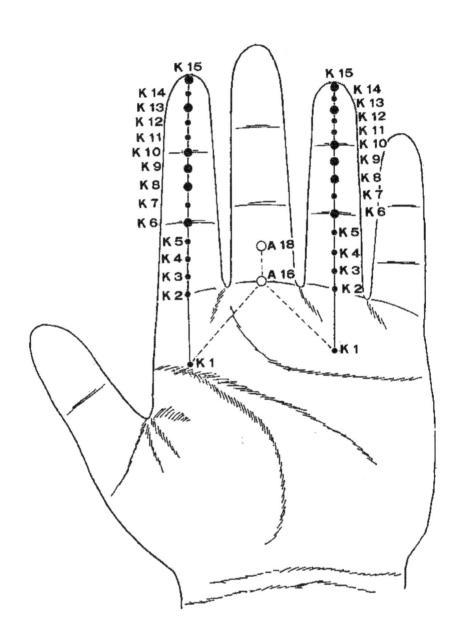

(12) L-Meridian (Triple Warmer)

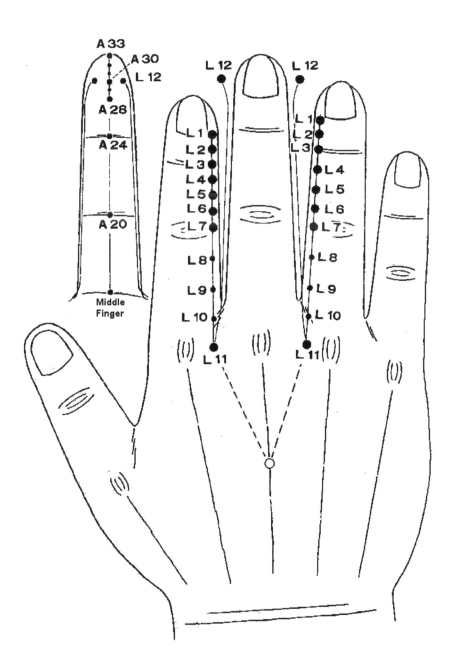

(13) M-Meridian (Gall Bladder)

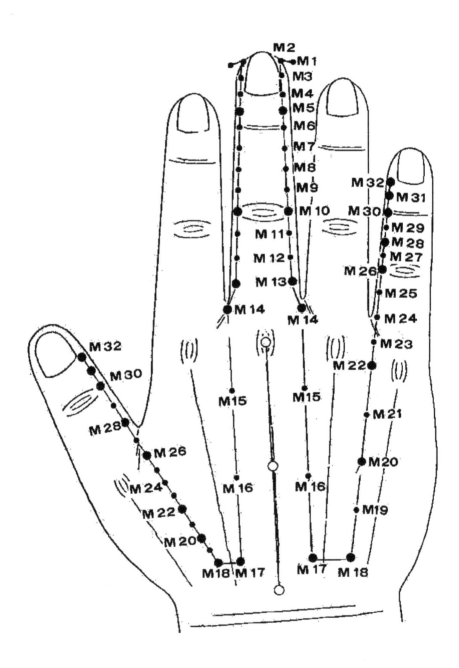

(14) N-Meridian (Liver)

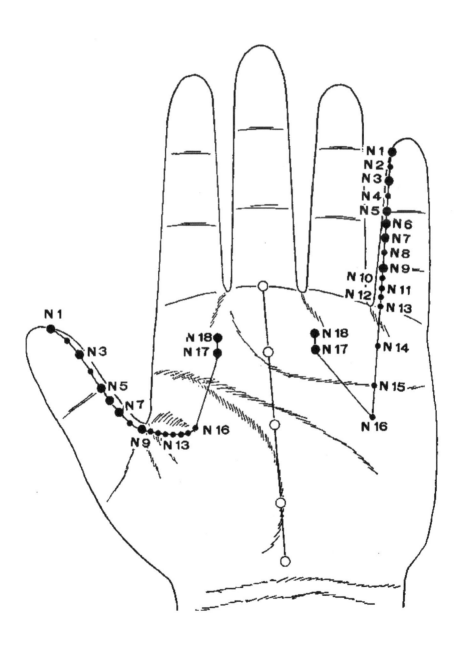

(15) Meridian Points on the Palm

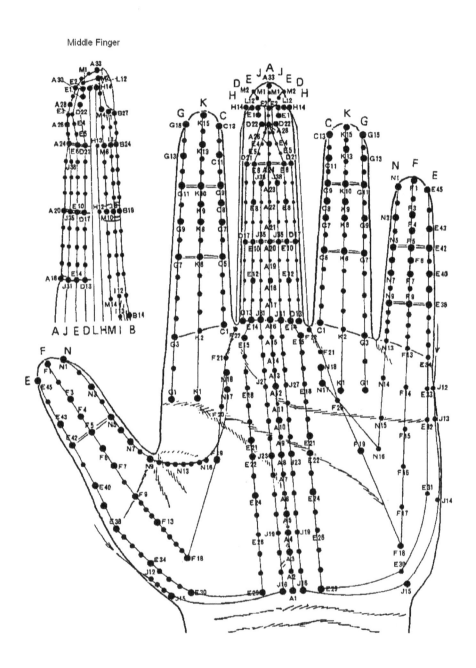

(16) Meridian Points on the Back of Hand

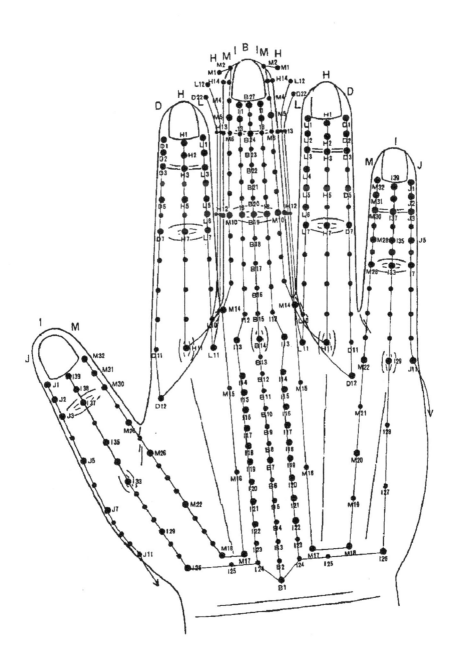

Index of Illnesses and Organs Affected

A
- Amenorrhea 29
- Asthma, Severe 65

B
- Backache:
    - Acute Lower Backache 47
    - Lower Backache 42
    - Lower Backache, had Three Surgeries 179
    - Lumbago, Bedridden for Forty Days 203
    - Lumbago, Fallen from Horse 174
    - Lumbago, Truck Driver's Chronic 183
    - Blood Pressure Lowered 11

C
- Cancer:
    - Liver Cancer, Spread 194
    - Liver Cancer, Terminal 198

Liver Cancer...4
Chest Pain after a Mastectomy.......................................70
Cold..157
Coma, Revived 79
Cramped Hand..89
Cramping Sole Instantly Relieved...............................102

E
　　　Ear Pain..134
　　　Eczema on the Scalp 132
　　　Eczema, Spread 1
　　　Esophageal Spasm 99
　　　Esophagitis, Chronic...162
　　　Eye, Infected..126

F
　　　Fallen from a Tree 109

G
　　　Graves' Disease..73

H
　　　Headache:
　　　Headache, Chronic..160
　　　　　　Migraine Headache 53
　　　Hearing Loss, Reversing 208
　　　Heart:
　　　　　　Heart Attack Stopped..23
　　　　　　Heart Disease...18
　　　　　　Heart Surgery (Pain)...77
　　　　　　Heart Was Beating Too Fast................................21
　　　　　　Holes in the Heart...9
　　　　　　Irregular Pulse...87

Hemorrhoid 113
Hepatitis 18
Hip, Injured 104

J

Jaw, Realigned 138

K

Kidney, Failure 119

L

Lump in the Back of Hand 45
Lump on the Chest 147

M

Menstrual Pain 124
Migraine Headache 53
Mute Cured 58
Myoma of the Uterus Cured 25

N

Nosebleed, Stop Instantly 129

O

Ovaritis 35

P

Pancreatitis 144
Paraplegia 92
Parkinson's Disease 83
Pimples 121
Prostate Patient Returns after Five Years 206
Prostate, Enlarged 187
PSA Count, Dropped More than 50% 191

R

Rectal Bleeding 39

Renal Failure 119
Restless Child Cured 115

S

Sciatica 170
Sciatica, Chronic 165
Seasick Young Man 150
Seasickness 154
Snoring 56
Spine, Dislocated 141
Stiffened Fingers Relaxed 96
Sweats Severely While Eating 62

T

Tonsillitis 136

W

Whiplash Injury 50

About the Author

Dr. Choong-Youl Oh was born in Seoul, Korea and spent his childhood in Japan and China, as the family moved with his Presbyterian missionary father. After the Second World War, the family returned to Korea. Dr. Oh studied pharmacy at Seoul National University and earned a BSc in pharmacy (Pharmacist Licence). With that degree, he taught chemistry at high-school level and at Yonsei University.

He came to Indiana University for graduate studies and acquired a master's degree in education and a PhD in instructional media and technology. During his US studies, he taught at Virginia State College and Indiana University. After graduation, he taught at the University of Alberta (Canada). He has now retired after forty years of teaching and is Professor Emeritus from the University of Alberta.

In pharmaceutical college, he studied biochemistry, pharmacology, and herbal medicine. More recently, he has studied pulse reading, acupuncture and Korean hand therapy, magnetic remedy, Genesen treatment, electronic beam therapy, and moxibustion treatment.

He has developed various effective alternative remedies. His primary interest is in the development of the treatment that gives immediate results, without pain, without side effects, and at low cost.

He has a patent from the US Patent Office in the treatment of neck and back injury and is applying for a patent for his innovative method in the treatment of an enlarged prostate gland.

Dr. Oh is an ardent Christian. In 1971, with the Rev. H. D. Kim, his former minister from Seoul, he founded the Edmonton Korean United Church, the first Korean-speaking church in Edmonton. As an elder, he has devoted himself to the church. He strongly believes that the most miraculous methods and formulas disclosed in this book were initiated by God: God's gift to us.

Dr. Oh is fluent in Korean, Japanese, and English. He earlier published this book in Korean as *Difficult Illness? No Problem* (2009).

Dr. Choong-Youl Oh can be reached at

Phone & Fax	1-780-487-3727
Phone	1-780-452-3603
Email	choong_oh@hotmail.com